S_pringer_ **M**_onographs_ in **M**_athematics_

For further volumes:
www.springer.com/series/3733

Luís Barreira

Dimension Theory of Hyperbolic Flows

Springer

Luís Barreira
Departamento de Matemática
Instituto Superior Técnico
Lisboa, Portugal

ISSN 1439-7382 Springer Monographs in Mathematics
ISBN 978-3-319-03392-1 ISBN 978-3-319-00548-5 (eBook)
DOI 10.1007/978-3-319-00548-5
Springer Cham Heidelberg New York Dordrecht London

Mathematics Subject Classification: 37C45, 37Dxx, 37Axx

Printed on acid-free paper

Springer is part of Springer Science+Business Media (www.springer.com)

To Claudia, for everything

Preface

The objective of this book is to provide a comprehensive exposition of the main results and main techniques of *dimension theory* and *multifractal analysis* of hyperbolic flows. This includes the discussion of some recent results in the area as well as some of its open problems.

The book is directed to researchers as well as graduate students specializing in dynamical systems who wish to have a sufficiently comprehensive view of the theory together with a working knowledge of its main techniques. The discussion of some open problems, perhaps somewhat biased towards my own interests, is included also with the hope that it may lead to further developments.

Over the last two decades, the dimension theory of dynamical systems has progressively developed into an independent and extremely active field of research. However, while the dimension theory and multifractal analysis for maps are very much developed, the corresponding theory for flows has experienced a steady although slower development. It should be emphasized that this is not because of lack of interest. For instance, geodesic flows and hyperbolic flows stand as cornerstones of the theory of dynamical systems. Sometimes a result for flows can be reduced to the case of maps, for example with the help of symbolic dynamics, but often it requires substantial changes or even new ideas. Because of this, many parts of the theory are either only sketched or are too technical for a wider audience. In this respect, the present monograph is intended to have a unifying and guiding role. Moreover, the text is self-contained and with the exception of some basic results in Chaps. 3 and 4, all the results in the book are included with detailed proofs.

On the other hand, there are topics that are not yet at a stage of development that makes it reasonable to include them in detail in a monograph of this nature, either because there are only partial results or because they require very specific techniques. This includes results for nonconformal flows, nonuniformly hyperbolic flows and flows modeled by countable symbolic dynamics. In such cases, I have instead provided a sufficient discussion with references to the relevant literature.

Lisbon, Portugal
June 2013

Luís Barreira

Contents

1 Introduction 1
1.1 Dimension Theory for Maps 1
1.2 Dimension Theory for Flows 5
1.3 Pointwise Dimension 6
1.4 Multifractal Analysis 8
1.5 Geodesic Flows 11
1.6 Variational Principles 13
1.7 Multidimensional Theory 13

Part I Basic Notions

2 Suspension Flows 19
2.1 Basic Notions and Cohomology 19
2.2 The Bowen–Walters Distance 25
2.3 Further Properties 29

3 Hyperbolic Flows 33
3.1 Basic Notions 33
3.2 Markov Systems 35
3.3 Symbolic Dynamics 36

4 Pressure and Dimension 39
4.1 Topological Pressure and Entropy 39
4.1.1 Basic Notions 39
4.1.2 Properties of the Pressure 41
4.1.3 The Case of Suspension Flows 42
4.2 BS-Dimension 43
4.3 Hausdorff and Box Dimensions 45
4.3.1 Dimension of Sets 45
4.3.2 Dimension of Measures 46

Part II Dimension Theory

5 Dimension of Hyperbolic Sets . . . 51
5.1 Dimensions Along Stable and Unstable Manifolds . . . 51
5.2 Formula for the Dimension . . . 58

6 Pointwise Dimension and Applications . . . 61
6.1 A Formula for the Pointwise Dimension . . . 61
6.2 Hausdorff Dimension and Ergodic Decompositions . . . 67
6.3 Measures of Maximal Dimension . . . 70

Part III Multifractal Analysis

7 Suspensions over Symbolic Dynamics . . . 81
7.1 Pointwise Dimension . . . 81
7.2 Multifractal Analysis . . . 83
7.3 Irregular Sets . . . 87
7.4 Entropy Spectra . . . 89

8 Multifractal Analysis of Hyperbolic Flows . . . 91
8.1 Suspensions over Expanding Maps . . . 91
8.2 Dimension Spectra of Hyperbolic Flows . . . 94
8.3 Entropy Spectra and Cohomology . . . 105

Part IV Variational Principles

9 Entropy Spectra . . . 111
9.1 A Conditional Variational Principle . . . 111
9.2 Analyticity of the Spectrum . . . 115
9.3 Examples . . . 119
9.3.1 Multifractal Spectra for the Local Entropies . . . 120
9.3.2 Multifractal Spectra for the Lyapunov Exponents . . . 121
9.3.3 Suspension Flows . . . 122
9.4 Multidimensional Spectra . . . 124

10 Multidimensional Spectra . . . 127
10.1 Multifractal Analysis . . . 127
10.2 Finer Structure . . . 134
10.3 Hyperbolic Flows: Analyticity of the Spectrum . . . 136

11 Dimension Spectra . . . 139
11.1 Future and Past . . . 139
11.2 Conditional Variational Principle . . . 141

References . . . 151

Index . . . 157

Chapter 1
Introduction

This introductory chapter gives an overview of the dimension theory and the multifractal analysis of dynamical systems, with emphasis on hyperbolic flows. Many of the results presented here are proved later on in the book. We also discuss topics that are not yet sufficiently well developed to include in the remaining chapters of a monograph of this nature. Finally, we include a discussion of open problems and suggestions for further developments.

1.1 Dimension Theory for Maps

The dimension theory of dynamical systems is an extremely active field of research. Its main objective is to measure the complexity from the dimensional point of view of the objects that remain invariant under the dynamics, such as the invariant sets and measures. We refer the reader to [3, 81] for detailed accounts of substantial parts of the dimension theory of dynamical systems (although these books make almost no reference to flows).

The objective of this book is to provide a comprehensive exposition of the main results and main techniques of dimension theory of hyperbolic flows. In this section we present some of these results. As a motivation, we start with a brief discussion of the corresponding theory for maps.

We first consider expanding maps. Let $g\colon M \to M$ be a differentiable map of a smooth manifold M and let $J \subset M$ be a compact g-invariant set. We say that J is a *repeller* of g and that g is an *expanding map* on J if there exist constants $c > 0$ and $\beta > 1$ such that

$$\|d_x g^n v\| \geq c\beta^n \|v\|$$

for every $n \in \mathbb{N}$, $x \in J$ and $v \in T_x M$. The map g is said to be *conformal* on J if $d_x g$ is a multiple of an isometry for every $x \in J$. We define a function $\varphi\colon J \to \mathbb{R}$ by

$$\varphi(x) = -\log\|d_x g\|.$$

L. Barreira, *Dimension Theory of Hyperbolic Flows*,
Springer Monographs in Mathematics, DOI 10.1007/978-3-319-00548-5_1,

The following result expresses the Hausdorff dimension $\dim_H J$ and the lower and upper box dimensions $\underline{\dim}_B J$ and $\overline{\dim}_B J$ (see Sect. 4.3 for the definitions) of a repeller J in terms of the topological pressure P (see Chap. 4).

Theorem 1.1 *If J is a repeller of a $C^{1+\delta}$ map g that is conformal on J, then*

$$\dim_H J = \underline{\dim}_B J = \overline{\dim}_B J = s,$$

where s is the unique real number such that

$$P(s\varphi) = 0. \tag{1.1}$$

Ruelle showed in [93] that $\dim_H J = s$. The equality between the Hausdorff and box dimensions is due to Falconer [39]. It is also shown in [93] that if μ is the equilibrium measure of $-s\varphi$, then

$$\dim_H J = \dim_H \mu, \tag{1.2}$$

where

$$\dim_H \mu = \inf\bigl\{\dim_H A : A \subset J \text{ and } \mu(J \setminus A) = 0\bigr\}.$$

Equation (1.1) was introduced by Bowen in [29] (in his study of quasi-circles) and is usually called *Bowen's equation*. It is also appropriate to call it the *Bowen–Ruelle equation*, taking into account the fundamental role played by the thermodynamic formalism developed by Ruelle in [91] (see also [92]) as well as his article [93]. Equation (1.1) establishes a connection between the thermodynamic formalism and dimension theory of dynamical systems. Since topological pressure and Hausdorff dimension are both Carathéodory characteristics (see [81]), the relation between the two is very natural. Moreover, equation (1.1) has a rather universal character: indeed, virtually all known equations used to compute or to estimate the dimension of an invariant set are particular cases of this equation or of an appropriate generalization (see [4] for a detailed discussion).

Now we consider hyperbolic sets. Let Λ be a hyperbolic set for a diffeomorphism f. We define functions $\varphi_s : \Lambda \to \mathbb{R}$ and $\varphi_u : \Lambda \to \mathbb{R}$ by

$$\varphi_s(x) = \log\|d_x f | E^s(x)\| \quad \text{and} \quad \varphi_u(x) = -\log\|d_x f | E^u(x)\|,$$

where $E^s(x)$ and $E^u(x)$ are respectively the stable and unstable subspaces at the point x. The set Λ is said to be *locally maximal* if there exists an open neighborhood U of Λ such that

$$\Lambda = \bigcap_{n \in \mathbb{Z}} f^n(U).$$

The following result is a version of Theorem 1.1 for hyperbolic sets.

Theorem 1.2 *If Λ is a locally maximal hyperbolic set for a C^1 surface diffeomorphism and* $\dim E^s(x) = \dim E^u(x) = 1$ *for every $x \in \Lambda$, then*

$$\dim_H \Lambda = \underline{\dim}_B \Lambda = \overline{\dim}_B \Lambda = t_s + t_u,$$

where t_s and t_u are the unique real numbers such that

$$P(t_s \varphi_s) = P(t_u \varphi_u) = 0.$$

It follows from work of McCluskey and Manning in [76] that $\dim_H \Lambda = t_s + t_u$. The equality between the Hausdorff and box dimensions is due to Takens [103] for C^2 diffeomorphisms and to Palis and Viana [79] for arbitrary C^1 diffeomorphisms. The result in Theorem 1.2 can be readily extended to the more general case of conformal diffeomorphisms. We recall that f is said to be *conformal* on a hyperbolic set Λ if the maps $d_x f | E^s(x)$ and $d_x f | E^u(x)$ are multiples of isometries for every $x \in \Lambda$. It happens that for conformal diffeomorphisms the product structure is a Lipschitz map with Lipschitz inverse (in general it is only a Hölder homeomorphism with Hölder inverse). This allows us to compute the dimension of a hyperbolic set by adding the dimensions along the stable and unstable manifolds.

Palis and Viana [79] established the continuous dependence of the dimension on the diffeomorphism. Mañé [72] obtained an even higher regularity. In higher-dimensional manifolds (and so in the nonconformal case) the Hausdorff dimension of a hyperbolic set may vary discontinuously. Examples were given by Pollicott and Weiss in [86] followed by Bonatti, Díaz, and Viana in [25]. Díaz and Viana [34] considered one-parameter families of diffeomorphisms on the 2-torus bifurcating from an Anosov map to a DA map and showed that for an open set of these families the Hausdorff and box dimensions of the nonwandering set are discontinuous across the bifurcation.

The study of the dimension of repellers and hyperbolic sets for nonconformal maps is much less developed than the corresponding theory for conformal maps. Some major difficulties include a clear separation between different Lyapunov directions, a small regularity of the associated distributions (that typically are only Hölder continuous), and the existence of number-theoretical properties forcing a variation of the Hausdorff dimension with respect to a certain typical value. As a result of this, in many situations only partial results have been obtained. For example, some results were obtained not for a particular transformation, but for Lebesgue almost all values of some parameter (although possibly without knowing what happens for a specific value of this parameter). Moreover, often only dimension estimates were obtained instead of a formula for the dimension of an invariant set.

This brief discussion of the difficulties encountered in the study of nonconformal maps motivates our first problem.

Problem 1.1 Develop a dimension theory for repellers and hyperbolic sets of nonconformal maps.

This is a very ambitious problem and in fact it should correspond to a large research program. Thus, it is reasonable to start with less general problems. This may involve, for example: to assuming that there is a clear separation between different Lyapunov exponents; to obtaining results for almost all values of some parameter and not for a specific transformation; or to obtaining sharp lower and upper bounds for the dimension instead of exact values. Several new phenomena occur in the study of nonconformal transformations. For example, in general the Hausdorff and box dimensions of a repeller do not coincide. An example was given by Pollicott and Weiss in [86], modifying a construction of Przytcki and Urbański in [88] depending on delicate number-theoretical properties.

Nevertheless, there exist many partial results towards a nonconformal theory, for several classes of repellers and hyperbolic sets, starting essentially with the seminal work of Douady and Oesterlé in [35]. In particular, Falconer [40] computed the Hausdorff dimension of a class of repellers for nonconformal maps (building on his former work [38]). Related results were obtained by Zhang in [110] and in the case of volume expanding maps by Gelfert in [49]. In another direction, Hu [59] computed the box dimension of a class of repellers for nonconformal maps leaving invariant a strong unstable foliation. Related results were obtained earlier by Bedford in [23] (see also [24]) for a class of self-similar sets that are graphs of continuous functions. In another direction, Falconer [37] studied a class of limit sets of geometric constructions obtained from the composition of affine transformations that are not necessarily conformal and he obtained a formula for the Hausdorff and box dimensions for Lebesgue almost all values in some parameter space (see also [102]). Related ideas were applied by Simon and Solomyak in [101] to compute the Hausdorff dimension of a class of solenoids in $\mathbb{R}^3$. Bothe [26] and then Simon [100] (also using his work in [99] for noninvertible transformations) studied earlier the dimension of solenoids. In particular, it is shown in [26] that under certain conditions on the dynamics the dimension is independent of the radial section (even though the holonomies are typically not Lipschitz). More recently, Hasselblatt and Schmeling conjectured in [55] (see also [54]) that, in spite of the difficulties due to the possible low regularity of the holonomies, the Hausdorff dimension of a hyperbolic set can be computed adding the dimensions along the stable and unstable manifolds. They prove this conjecture for a class of solenoids. The ideas developed in all these works should play an important role in the study of Problem 1.1.

There also exist some related results for nonuniformly expanding maps. In particular, Horita and Viana [57] and Dysman [36] studied abstract models, called maps with holes, which include examples of nonuniformly expanding repellers. In [58] Horita and Viana considered nonuniformly expanding repellers emerging from a perturbation of an Anosov diffeomorphism of the 3-torus through a Hopf bifurcation. Finally, we mention some related work in the case of nonuniformly hyperbolic invariant sets. Hirayama [56] obtained an upper bound for the Hausdorff dimension of the stable set of the set of typical points for a hyperbolic measure. Fan, Jiang and Wu [43] studied the dimension of the maximal invariant set of an asymptotically nonhyperbolic family. Urbánski and Wolf [104] considered horseshoe maps that are uniformly hyperbolic except at a parabolic point, in particular establishing a dimension formula for the horseshoe.

In connection with identity (1.2) another interesting problem is the following.

Problem 1.2 Given a repeller for a nonconformal map, find whether there exists an invariant measures of full dimension.

Identity (1.2) is due to Ruelle [93] and follows from the equivalence between μ and the s-dimensional Hausdorff measure on J. The existence of an ergodic measure of full dimension on a repeller of a C^1 conformal map was established by Gatzouras and Peres in [48]. The situation is much more complicated in the case of nonconformal maps, and there exist only some partial results. In particular, it is shown in [48] that repellers of some maps of product type also have ergodic measures of full dimension. For piecewise linear maps, Gatzouras and Lalley [47] showed earlier that certain invariant sets, corresponding to full shifts in the symbolic dynamics, carry an ergodic measure of full dimension. Kenyon and Peres [64] obtained the same result for linear maps and arbitrary compact invariant sets. Bedford and Urbański considered a particular class of self-affine sets in [24] and obtained conditions for the existence of a measure of full dimension. Related ideas appeared earlier in work of Bedford [22] and McMullen [77]. More recently, Yayama [107] considered general Sierpiński carpets modeled by arbitrary topological Markov chains and Luzia [70, 71] considered expanding triangular maps of the 2-torus.

1.2 Dimension Theory for Flows

Now we turn to the case of flows. To a large extent the theory is analogous to the theory for maps. Let Φ be a $C^{1+\delta}$ flow with a locally maximal hyperbolic set Λ such that $\Phi|\Lambda$ is conformal and topologically mixing (see Sect. 5.1). Let also $V^s(x)$ and $V^u(x)$ be the families of local stable and unstable manifolds (see Sect. 3.1). The following result of Pesin and Sadovskaya in [82] expresses the dimensions of the sets $V^s(x)\cap\Lambda$ and $V^u(x)\cap\Lambda$ in terms of the topological pressure (see Theorem 5.1). We define functions $\zeta_s, \zeta_u\colon \Lambda\to\mathbb{R}$ by

$$\zeta_s(x)=\frac{\partial}{\partial t}\log\|d_x\varphi_t|E^s(x)\|\Big|_{t=0}$$

and

$$\zeta_u(x)=\frac{\partial}{\partial t}\log\|d_x\varphi_t|E^u(x)\|\Big|_{t=0}.$$

Theorem 1.3 *Let Φ be a $C^{1+\delta}$ flow with a locally maximal hyperbolic set Λ such that $\Phi|\Lambda$ is conformal and topologically mixing. Then*

$$\dim_H(V^s(x)\cap\Lambda)=\underline{\dim}_B(V^s(x)\cap\Lambda)=\overline{\dim}_B(V^s(x)\cap\Lambda)=t_s$$

and

$$\dim_H(V^u(x)\cap\Lambda)=\underline{\dim}_B(V^u(x)\cap\Lambda)=\overline{\dim}_B(V^u(x)\cap\Lambda)=t_u,$$

where t_s and t_u are the unique real numbers such that

$$P_{\Phi|\Lambda}(t_s\zeta_s) = P_{\Phi|\Lambda}(-t_u\zeta_u) = 0.$$

It is also shown in [82] that

$$\dim_H \Lambda = \underline{\dim}_B \Lambda = \overline{\dim}_B \Lambda = t_s + t_u + 1$$

(see Theorem 5.2). This is a version of Theorem 1.2 for flows.

We also formulate a version of Problem 1.1 for hyperbolic flows.

Problem 1.3 Develop a dimension theory for hyperbolic sets of nonconformal flows.

Similar comments to those in the case of maps apply to Problem 1.3. However, the theory is largely untouched ground. In particular, to the best of our knowledge, the first lower and upper bounds for the dimensions along the stable and unstable manifolds of a hyperbolic set, for a nonconformal flow, appear for the first time in this book (see (5.18)).

1.3 Pointwise Dimension

In the theory of dynamical systems, each *global* quantity can often be "constructed" with the help of a certain *local* quantity. Two examples are the Kolmogorov–Sinai entropy and the Hausdorff dimension, which are quantities of a global nature. They can be built (in a rigorous mathematical sense) using respectively the local entropy and the pointwise dimension. More precisely, in the case of the entropy this goes back to the classical Shannon–McMillan–Breiman theorem: the Kolmogorov–Sinai entropy is obtained by integrating the local entropy. On the other hand, the Hausdorff dimension of a measure is given by the essential supremum of the lower pointwise dimension. More precisely, given a measure μ in a set $\Lambda \subset \mathbb{R}^m$, we have

$$\dim_H \mu = \text{ess sup}\left\{\liminf_{r\to 0} \frac{\log \mu(B(x,r))}{\log r} : x \in \Lambda\right\},$$

where $B(x,r)$ is the ball of radius r centered at x, with the essential supremum taken with respect to μ. In particular, if there exists a real number d such that

$$\lim_{r\to 0} \frac{\log \mu(B(x,r))}{\log r} = d \tag{1.3}$$

for μ-almost every $x \in \Lambda$, then $\dim_H \mu = d$. This criterion was established by Young in [108]. The limit in (1.3), if it exists, is called the *pointwise dimension* of μ at x. Let μ be a finite measure with compact support that is invariant under a $C^{1+\delta}$ diffeomorphism f. It follows from work of Ledrappier and Young in [67, 68] and work of Barreira, Pesin and Schmeling in [11] that if the measure μ is hyperbolic (that is, if all Lyapunov exponents are nonzero μ-almost everywhere), then the pointwise dimension exists almost everywhere (see [10] for details). In the two-dimensional case this statement was established by Young in [108].

Let $f: M \to M$ be a $C^{1+\delta}$ surface diffeomorphism. For each $x \in M$ and $v \in T_x M$, we consider the Lyapunov exponent

$$\lambda(x, v) = \limsup_{n\to+\infty} \frac{1}{n} \log\|d_x f^n v\|.$$

Let μ be an f-invariant probability measure on M. We say that μ is *hyperbolic* if $\lambda(x, v) \neq 0$ for μ-almost every $x \in M$ and every $v \neq 0$. When μ is of saddle type, that is, when the function $T_x M \setminus \{0\} \mapsto \lambda(x, v)$ takes exactly one positive value $\lambda_u(x)$ and one negative value $\lambda_s(x)$, for μ-almost every $x \in M$, we define

$$\lambda_u(\mu) = \int_M \lambda_u \, d\mu \quad \text{and} \quad \lambda_s(\mu) = \int_M \lambda_s \, d\mu.$$

Moreover, we denote by $h_\mu(f)$ the entropy of f with respect to μ. By work of Brin and Katok in [32], the limit

$$h_\mu(x) = \lim_{\varepsilon\to 0} \lim_{n\to\infty} -\frac{1}{n} \log \mu\left(\bigcap_{k=0}^{n-1} f^{-k} B(f^k(x), \varepsilon)\right)$$

exists for μ-almost every $x \in M$ and

$$h_\mu(f) = \int_M h_\mu(x) \, d\mu.$$

The number $h_\mu(x)$ is called the *local entropy* of μ at the point x.

The following result was established by Young in [108].

Theorem 1.4 *Let f be a $C^{1+\delta}$ diffeomorphism. If μ is an ergodic f-invariant measure, then*

$$\dim_H \mu = h_\mu(f)\left(\frac{1}{\lambda_u(\mu)} - \frac{1}{\lambda_s(\mu)}\right). \tag{1.4}$$

Barreira and Wolf [20] considered measures sitting on a hyperbolic set that are not necessarily ergodic and established an explicit formula for the pointwise dimension, which is a local version of identity (1.4).

Theorem 1.5 *Let f be a $C^{1+\delta}$ surface diffeomorphism with a locally maximal hyperbolic set Λ and let μ be an f-invariant probability measure on Λ. For μ-almost every $x \in \Lambda$, we have*

$$\lim_{r\to 0} \frac{\log \mu(B(x, r))}{\log r} = h_\mu(x)\left(\frac{1}{\lambda_u(x)} - \frac{1}{\lambda_s(x)}\right).$$

The novelty of the approach in [20] is not Theorem 1.5 itself, but instead the elementary method of proof. Indeed, the result also follows from work of Ledrappier and Young in [68], although with a rather involved proof in the general context of nonuniform hyperbolicity (see [10] for details).

In [21], Barreira and Wolf established an analogous formula for conformal hyperbolic flows (see Theorem 6.1).

Theorem 1.6 *Let Φ be a $C^{1+\delta}$ flow with a locally maximal hyperbolic set Λ such that $\Phi|\Lambda$ is conformal and let μ be a Φ-invariant probability measure on Λ. For μ-almost every $x \in \Lambda$, we have*

$$\lim_{r\to 0} \frac{\log \mu(B(x,r))}{\log r} = h_\mu(x)\left(\frac{1}{\lambda_u(x)} - \frac{1}{\lambda_s(x)}\right) + 1. \tag{1.5}$$

In [82], Pesin and Sadovskaya first established identity (1.5) in the special case of equilibrium measures for a Hölder continuous function (we note that these measures are ergodic and have a local product structure). Identity (1.5) can be used to describe how the Hausdorff dimension $\dim_H \mu$ behaves under an ergodic decomposition. We recall that an ergodic decomposition of a measure μ can be identified with a probability measure τ in the metrizable space of Φ-invariant probability measures on Λ such that the subset of ergodic measures has full τ-measure (see Sect. 4.3). Namely, for any ergodic decomposition of μ we have

$$\dim_H \mu = \operatorname{ess\,sup}_\nu \dim_H \nu,$$

with the essential supremum taken with respect to τ (see Theorem 6.3).

The discussion in the case of maps motivates the following problem.

Problem 1.4 Establish identity (1.5) for an arbitrary hyperbolic measure.

We emphasize that this is a very ambitious problem. The main difficulty seems to be that it should be necessary to develop appropriate tools in the context of a nonuniform hyperbolicity theory for flows. Nevertheless, it is reasonable to conjecture that identity (1.5) indeed holds for any hyperbolic measure.

1.4 Multifractal Analysis

The multifractal analysis of dynamical systems can be considered a subfield of the dimension theory of dynamical systems. Roughly speaking, multifractal analysis studies the complexity of the level sets of invariant local quantities obtained from a dynamical system. For example, one can consider Birkhoff averages, Lyapunov exponents, pointwise dimensions and local entropies. These functions are typically only measurable and thus their level sets are rarely manifolds. Hence, in order to measure their complexity it is appropriate to use quantities such as the topological entropy and the Hausdorff dimension.

The concept of multifractal analysis was suggested by Halsey, Jensen, Kadanoff, Procaccia and Shraiman in [51]. The first rigorous approach is due to Collet, Lebowitz and Porzio in [33] for a class of measures invariant under one-dimensional Markov maps. In [69], Lopes considered the measure of maximal entropy for hyperbolic Julia sets, and in [89], Rand studied Gibbs measures for a class of repellers. We refer the reader to the books [3, 81] for detailed accounts of substantial parts of the theory.

We briefly describe the main elements of multifractal analysis. Let $T\colon X \to X$ be a continuous map of a compact metric space and let $g\colon X \to \mathbb{R}$ be a continuous function. For each $\alpha \in \mathbb{R}$, let

$$K_\alpha = \left\{ x \in X : \lim_{n\to\infty} \frac{1}{n} \sum_{i=0}^{n} g(T^i(x)) = \alpha \right\}. \tag{1.6}$$

We also consider the set

$$K = \left\{ x \in X : \liminf_{n\to\infty} \frac{1}{n} \sum_{i=0}^{n} g(T^i(x)) < \limsup_{n\to\infty} \frac{1}{n} \sum_{i=0}^{n} g(T^i(x)) \right\}. \tag{1.7}$$

Clearly,

$$X = K \cup \bigcup_{\alpha\in\mathbb{R}} K_\alpha. \tag{1.8}$$

This union is formed by pairwise disjoint T-invariant sets. It is called a *multifractal decomposition* of X. For each $\alpha \in \mathbb{R}$, let

$$\mathcal{D}(\alpha) = \dim_H K_\alpha.$$

The function $\mathcal{D}$ is called the *dimension spectrum for the Birkhoff averages of* g.

By Birkhoff's ergodic theorem, if μ is an ergodic T-invariant finite measure on X, and $\alpha = \int_X g\,d\mu/\mu(X)$, then $\mu(K_\alpha) = \mu(X)$. That is, there exists a set K_α in the multifractal decomposition of full μ-measure. However, the other sets in the multifractal decomposition need not be empty. In fact, for several classes of hyperbolic dynamical systems (for example, a topological Markov chain, an expanding map or a hyperbolic diffeomorphism) and certain functions g (for example, Hölder continuous functions that are not cohomologous to a constant), it was proved that:

1. the set $\{\alpha \in \mathbb{R} : K_\alpha \neq \varnothing\}$ is an interval;
2. the function $\mathcal{D}$ is analytic and strictly convex;
3. the set K is everywhere dense and $\dim_H K = \dim_H X$.

In particular, the multifractal decomposition in (1.8) is often composed of an uncountable number of dense T-invariant sets of positive Hausdorff dimension. For example, for repellers and hyperbolic sets for $C^{1+\delta}$ conformal maps, Pesin and

Weiss [83, 84] obtained a multifractal analysis of the dimension spectrum. We refer the reader to [3, 4] for details and further references.

As in the case of the dimension theory of nonconformal maps, the study of the corresponding multifractal analysis is still at an early stage of development and the following is a challenging problem.

Problem 1.5 Obtain a multifractal analysis for repellers and hyperbolic sets for nonconformal maps.

We mention briefly some works containing partial results towards a solution of this problem. Feng and Lau [46] and Feng [44, 45] studied products of nonnegative matrices and their thermodynamic properties. Jordan and Simon [61] established formulas for the dimension spectra of almost all self-affine maps in the plane (we note that their results generalize to any dimension). In [8], Barreira and Gelfert considered repellers of nonconformal maps satisfying a certain cone condition and obtained a multifractal analysis for the topological entropy of the level sets of the Lyapunov exponents.

Another challenging problem concerns nonuniformly hyperbolic maps. We emphasize that in this case even the conformal case is at an early stage of development.

Problem 1.6 Obtain a multifractal analysis for nonuniformly hyperbolic maps.

The intrinsic difficulty of this problem is not strictly related to the general class of dynamics under consideration. Indeed, there are essential differences between the thermodynamic formalisms for uniformly hyperbolic and nonuniformly hyperbolic dynamics. Due to the important role played by the thermodynamic formalism, this can be seen as the main reason behind important differences between the dimension theories for uniformly hyperbolic and nonuniformly hyperbolic dynamics.

We also mention some works related to Problem 1.6. We first observe that for uniformly hyperbolic systems and their codings by finite topological Markov chains, the dimension and entropy spectra of an equilibrium measure has bounded domain and is analytic. In strong contrast, in the case of nonuniformly hyperbolic systems and countable topological Markov chains the spectrum may have unbounded domain and need not be analytic. In [87], Pollicott and Weiss considered the Gauss map and the Manneville–Pomeau transformation. Related results were obtained by Yuri in [109]. In [73–75], Mauldin and Urbański developed the theory of infinite conformal iterated function systems, studying in particular the Hausdorff dimension of the limit set (see also [52]). Related results were obtained by Nakaishi in [78]. In [66], Kesseböhmer and Stratmann established a detailed multifractal analysis for Stern–Brocot intervals, continued fractions and certain Diophantine growth rates, building on their former work [65]. We refer to [85] for results concerning Farey trees and multifractal analysis. In [60], Iommi obtained a multifractal analysis for countable topological Markov chains. He uses the Gurevich pressure introduced by Sarig in [94] (building on former work of Gurevich [50] on the notion of topological entropy for countable Markov chains).

One can also obtain a multifractal analysis for a class of hyperbolic flows and for suspension flows over topological Markov chains. In the multifractal analysis of a flow $\Phi = \{\varphi_t\}_{t\in\mathbb{R}}$ in X, the sets K_α and K in (1.6) and (1.7) are replaced respectively by

$$K_\alpha = \left\{x \in X : \lim_{t\to\infty} \frac{1}{t}\int_0^t g(\varphi_\tau(x))\,d\tau = \alpha\right\}$$

and

$$K = \left\{x \in X : \liminf_{t\to\infty} \frac{1}{t}\int_0^t g(\varphi_\tau(x))\,d\tau < \limsup_{t\to\infty} \frac{1}{t}\int_0^t g(\varphi_\tau(x))\,d\tau\right\}.$$

In particular:

1. Pesin and Sadovskaya [82] obtained a multifractal analysis of the dimension spectrum for the pointwise dimensions of a Gibbs measure on a locally maximal hyperbolic set for a conformal flow (see Theorem 8.3);
2. Barreira and Saussol [12] obtained a multifractal analysis of the entropy spectrum for the Birkhoff averages of a Hölder continuous function on a locally maximal hyperbolic set (see Theorem 8.4).

The main idea of the proofs is to use Markov systems and the associated symbolic dynamics developed by Bowen [27] and Ratner [90] to reduce the setup to the case of maps. This is done using suspension flows over topological Markov chains, obtained from a Markov system, and a careful analysis of the relation between the cohomology for the flow and the cohomology for the map in the base. Later on in the book, we describe more general results with proofs that do not use Markov systems and the associated symbolic dynamics.

The following are versions of Problems 1.5 and 1.6 for flows.

Problem 1.7 Obtain a multifractal analysis for repellers and hyperbolic sets for nonconformal flows.

Problem 1.8 Obtain a multifractal analysis for nonuniformly hyperbolic flows.

In [9], Barreira and Iommi considered suspension flows over a countable topological Markov chain, building also on work of Savchenko [95] on the notion of topological entropy.

1.5 Geodesic Flows

In this section we discuss an application of the multifractal analysis for flows to geodesic flows on compact surfaces of negative curvature.

Consider a compact orientable Riemannian surface M with (sectional) curvature K. The Gauss–Bonnet theorem says that

$$\int_M K\,d\lambda_M = 2\pi\chi(M), \tag{1.9}$$

where λ_M is the volume in M and $\chi(M)$ is the Euler characteristic of M. Let $\Phi = \{\varphi_t\}_{t\in\mathbb{R}}$ be the geodesic flow in the unit tangent bundle SM. It preserves the normalized Liouville measure λ_{SM} in SM, induced from the volume in M. By Birkhoff's ergodic theorem, the limit

$$\kappa(x) = \lim_{t\to\infty} \frac{1}{t}\int_0^t K(\varphi_s(x))\,ds \tag{1.10}$$

exists for λ_{SM}-almost every $x \in SM$. It follows from (1.9) and (1.10) that

$$\int_{SM} \kappa\,d\lambda_{SM} = \int_M K\,d\lambda_M = 2\pi\chi(M). \tag{1.11}$$

Now let us assume that M has strictly negative curvature. In this case M has genus at least 2. The geodesic flow is ergodic and hence, in addition to (1.11), we have

$$\kappa(x) = \int_M K\,d\lambda_M = 2\pi\chi(M) \tag{1.12}$$

for λ_{SM}-almost every $x \in SM$. More generally, identity (1.12) holds almost everywhere with respect to any invariant probability measure. However, the level sets

$$SM_\alpha = \{x \in SM : \kappa(x) = \alpha\}$$

may still be nonempty for some values of α. The following result was established by Barreira and Saussol in [12].

Theorem 1.7 *Given a compact orientable surface M with $\chi(M) < 0$, for each metric g in an open set of C^3 metrics in M of strictly negative curvature, there exists an open interval I_g containing $2\pi\chi(M)$ such that $SM_\alpha \subset SM$ is a nonempty proper dense subset with $h(\Phi|SM_\alpha) > 0$ for every $\alpha \in I_g$.*

We have

$$SM = N \cup \bigcup_\alpha SM_\alpha,$$

where

$$N = \left\{x \in SM : \liminf_{t\to\infty} \frac{1}{t}\int_0^t K(\varphi_s(x))\,ds < \limsup_{t\to\infty} \frac{1}{t}\int_0^t K(\varphi_s(x))\,ds\right\}$$

and the union is composed of pairwise disjoint sets. By Birkhoff's ergodic theorem, the set N has zero measure with respect to any invariant measure. This strongly contrasts with the following result also obtained in [12] (using ideas in [17]).

Theorem 1.8 *Given a compact orientable surface M with $\chi(M) < 0$, for each metric g in an open set of C^3 metrics in M of strictly negative curvature, the set $N \subset SM$ is a nonempty proper dense subset with $h(\Phi|N) = h(\Phi)$.*

1.6 Variational Principles

In this section we describe another approach to the multifractal analysis of entropy spectra, based on what we call a conditional variations principle. For simplicity of the presentation, in order to avoid introducing extra material at this point, we consider again the setup of Sect. 1.5 (the general case is considered in Sect. 1.7).

Let $\Phi = \{\varphi_t\}_{t\in\mathbb{R}}$ be a geodesic flow in the unit tangent bundle SM. For each $\alpha \in \mathbb{R}$, let

$$\mathcal{E}(\alpha) = h(\Phi|SM_\alpha)$$

be the topological entropy of Φ on the set SM_α. The function $\mathcal{E}$ is called the *entropy spectrum*. In many works of multifractal analysis the function $\mathcal{E}$ is described in terms of a Legendre transform involving the topological pressure. A conditional variational principle provides an alternative description.

The following result of Barreira and Saussol in [15] is a conditional variational principle for the spectrum $\mathcal{E}$. Let $h_\mu(\Phi)$ be the entropy of the geodesic flow with respect to a measure $\mu \in \mathcal{M}$, where $\mathcal{M}$ is the set of all Φ-invariant probability measures on SM.

Theorem 1.9 *For a compact orientable surface M and a metric of strictly negative curvature on M, for each*

$$\alpha \in \operatorname{int}\left\{\int_{SM} K\,d\mu : \mu \in \mathcal{M}\right\},$$

we have

$$\mathcal{E}(\alpha) = \max\left\{h_\mu(\Phi) : \int_{SM} K\,d\mu = \alpha \text{ and } \mu \in \mathcal{M}\right\}.$$

Theorem 1.9 is a particular case of Theorem 9.1.

1.7 Multidimensional Theory

In this section we consider multidimensional versions of entropy and dimension spectra for a flow Φ with upper semicontinuous entropy $\mu \mapsto h_\mu(\Phi)$ and we describe a conditional variational principle for these spectra. This allows us to study simultaneously the level sets of several local quantities, instead of only one as in Sect. 1.4.

For example, the entropy of a C^1 flow with a hyperbolic set is upper semicontinuous. More generally, the entropy of any expansive flow is upper semicontinuous. On the other hand, there are many transformations without a hyperbolic set (and not satisfying specification) for which the entropy is still upper semicontinuous. For example, all β-shifts are expansive and thus, the entropy is upper semicontinuous (see [63] for details), but for β in a residual set of full Lebesgue measure (although the complement has full Hausdorff dimension) the corresponding β-shift does not satisfy specification (see [96]). This motivates establishing results not only for flows with a hyperbolic set but more generally for flows with upper semicontinuous entropy. Moreover, we consider functions with a unique equilibrium measure. It follows from work of Walters [105] that for each β-shift any Lipschitz function has a unique equilibrium measure. We recall that for topologically mixing hyperbolic flows each Hölder continuous function has a unique equilibrium measure.

Now we consider a continuous flow Φ in a compact metric space X. Given continuous functions $a_1, a_2 \colon X \to \mathbb{R}$, we consider the level sets of Birkhoff averages

$$K_{\alpha_1,\alpha_2} = \left\{ x \in X : \lim_{t\to\infty} \frac{1}{t} \int_0^t a_i(\varphi_s(x))\, ds = \alpha_i \text{ for } i = 1,2 \right\}.$$

The associated *entropy spectrum* is defined by

$$\mathcal{E}(\alpha_1,\alpha_2) = h(\Phi | K_{\alpha_1,\alpha_2}).$$

We also consider the set

$$\mathcal{P} = \left\{ \left(\int_X a_1\, d\mu, \int_X a_2\, d\mu \right) : \mu \in \mathcal{M} \right\},$$

where $\mathcal{M}$ is the family of all Φ-invariant probability measures on X. The following is a conditional variational principle for the spectrum $\mathcal{E}$.

Theorem 1.10 *Assume that the map $\mu \mapsto h_\mu(\Phi)$ is upper semicontinuous and that for each $c_1, c_2 \in \mathbb{R}$ the function $c_1 a_1 + c_2 a_2$ has a unique equilibrium measure. Then for each $(\alpha_1,\alpha_2) \in \operatorname{int}\mathcal{P}$, we have*

$$\mathcal{E}(\alpha_1,\alpha_2) = \max \left\{ h_\mu(\Phi) : \left(\int_X a_1\, d\mu, \int_X a_2\, d\mu \right) = (\alpha_1,\alpha_2) \text{ and } \mu \in \mathcal{M} \right\}$$

and there exists an ergodic measure $\mu \in \mathcal{M}$ with $\mu(K_{\alpha_1,\alpha_2}) = 1$ such that

$$h_\mu(\Phi) = \mathcal{E}(\alpha_1,\alpha_2) \quad \text{and} \quad \left(\int_X a_1\, d\mu, \int_X a_2\, d\mu \right) = (\alpha_1,\alpha_2).$$

Theorem 1.10 is a particular case of Theorem 10.1 due to Barreira and Doutor [6]. In the case when Φ is a hyperbolic flow, the statement in Theorem 1.10 was first established by Barreira and Saussol in [15]. This study revealed new nontrivial phenomena absent in one-dimensional multifractal analysis. In particular, while the domain of a one-dimensional spectrum is always an interval, for multidimensional spectra it may not be convex and may have empty interior, although still containing uncountably many points. Moreover, the interior of the domain of a multidimensional spectrum may have more than one connected component. We refer to [16] for a detailed discussion.

The proof of Theorem 1.10 is based on techniques developed by Barreira, Saussol and Schmeling in [16] and Barreira and Saussol in [15]. We emphasize that this approach deals directly with the flows and in particular it does not require Markov systems.

Part I
Basic Notions

This part is of an introductory nature and serves as a reference for the remaining chapters. We recall in a pragmatic manner all the necessary notions and results from hyperbolic dynamics, the thermodynamic formalism and dimension theory that are needed in the book. In Chap. 2 we consider suspension flows, the notion of cohomology and the Bowen–Walters distance. Suspension flows serve as models for hyperbolic flows which are introduced in Chap. 3. Here we also recall the notion of a Markov system and we describe how it can be used to associate a symbolic dynamics to any locally maximal hyperbolic set. In Chap. 4 we recall all the necessary notions and results from the thermodynamic formalism and dimension theory. This includes the notions of topological pressure, BS-dimension, lower and upper box dimensions and pointwise dimension.

Chapter 2
Suspension Flows

In this chapter we present several basic notions and results regarding suspension flows, as a preparation for many developments in later chapters. We note that any smooth flow with a hyperbolic set gives rise to a suspension flow (see Chap. 3). In particular, we present the notions of cohomology and of Bowen–Walters distance. It happens that one can often describe the properties of a suspension flow in terms of corresponding properties in the base. This relation is considered in this chapter for the notion of cohomology. Several other relations of a similar type will be considered later in the book.

2.1 Basic Notions and Cohomology

We first introduce the notion of a suspension flow. Let $T: X \to X$ be a homeomorphism of a compact metric space and let $\tau: X \to (0, \infty)$ be a Lipschitz function. Consider the space

$$Z = \{(x, s) \in X \times \mathbb{R} : 0 \le s \le \tau(x)\},$$

and let Y be the set obtained from Z by identifying the points $(x, \tau(x))$ and $(T(x), 0)$ for each $x \in X$. One can introduce in a natural way a topology on Z and thus also on Y (obtained from the product topology on $X \times \mathbb{R}$), with respect to which Y is a compact topological space. This topology is induced by a certain distance introduced by Bowen and Walters in [31] (see Sect. 2.2).

Definition 2.1 The *suspension flow over T with height function τ* is the flow $\Psi = \{\psi_t\}_{t\in\mathbb{R}}$ in Y with the maps $\psi_t: Y \to Y$ defined by

$$\psi_t(x, s) = (x, s + t) \tag{2.1}$$

(see Fig. 2.1).

L. Barreira, *Dimension Theory of Hyperbolic Flows*,
Springer Monographs in Mathematics, DOI 10.1007/978-3-319-00548-5_2,

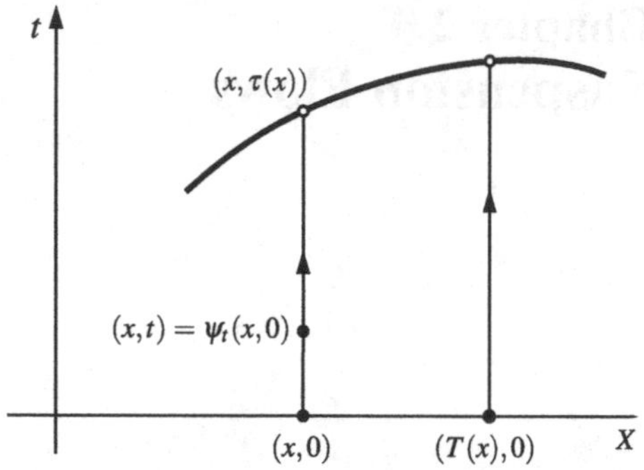

Fig. 2.1 A suspension flow $\Psi = \{\psi_t\}_{t\in\mathbb{R}}$ over T

We note that any suspension flow is indeed a flow, that is, $\psi_0 = \mathrm{id}$ and

$$\psi_t \circ \psi_s = \psi_{t+s} \quad \text{for} \quad t, s \in \mathbb{R}.$$

The set X is called the *base* of the suspension flow. We extend τ to a function $\tau: Y \to \mathbb{R}$ by

$$\tau(y) = \min\{t > 0 : \psi_t(y) \in X \times \{0\}\},$$

and T to a map $T: Y \to X \times \{0\}$ by $T(y) = \psi_{\tau(y)}(y)$ (since there is no danger of confusion, we continue to use the symbols τ and T for the extensions).

Now we introduce the notion of cohomology for flows and maps.

Definition 2.2 (Notion of Cohomology)

1. A function $g: Y \to \mathbb{R}$ is said to be Ψ*-cohomologous* to a function $h: Y \to \mathbb{R}$ in a set $A \subset Y$ if there exists a bounded measurable function $q: Y \to \mathbb{R}$ such that

$$g(x) - h(x) = \lim_{t\to 0} \frac{q(\psi_t(x)) - q(x)}{t} \tag{2.2}$$

for every $x \in A$.
2. A function $G: Y \to \mathbb{R}$ is said to be T*-cohomologous* to a function $H: Y \to \mathbb{R}$ in a set $A \subset Y$ if there exists a bounded measurable function $q: Y \to \mathbb{R}$ such that

$$G(x) - H(x) = q(T(x)) - q(x)$$

for every $x \in A$.

One can easily verify that the two notions of cohomology in Definition 2.2 are equivalence relations. We refer to the corresponding equivalence classes as *cohomology classes*.

Now we show that the notion of cohomology for a suspension flow can be described in terms of the notion of cohomology for the map in the base. This observation will be very useful in some of the proofs. Given a continuous function

$g: Y \to \mathbb{R}$, we define a new function $I_g: Y \to \mathbb{R}$ by

$$I_g(y) = \int_0^{\tau(y)} g(\psi_s(y))\,ds. \tag{2.3}$$

Theorem 2.1 *If Ψ is a suspension flow over $T: X \to X$ and $g, h: Y \to \mathbb{R}$ are continuous functions, then the following properties are equivalent:*

1. *g is Ψ-cohomologous to h in Y, with*

$$g(y) - h(y) = \lim_{t\to 0} \frac{q(\psi_t(y)) - q(y)}{t} \quad \textit{for} \quad y \in Y;$$

2. *I_g is T-cohomologous to I_h in Y, with*

$$I_g(y) - I_h(y) = q(T(y)) - q(y) \quad \textit{for} \quad y \in Y; \tag{2.4}$$

3. *I_g is T-cohomologous to I_h in $X \times \{0\}$, with*

$$I_g(y) - I_h(y) = q(T(y)) - q(y) \quad \textit{for} \quad y \in X \times \{0\}.$$

Proof We first assume that g is Ψ-cohomologous to h in Y. For each $y \in Y$, we have

$$\begin{aligned}
I_g(y) - I_h(y) &= \int_0^{\tau(y)} \lim_{t\to 0} \frac{q(\psi_t(\psi_s(y))) - q(\psi_s(y))}{t}\,ds \\
&= \lim_{t\to 0} \frac{1}{t}\left(\int_t^{\tau(y)+t} q(\psi_s(y))\,ds - \int_0^{\tau(y)} q(\psi_s(y))\,ds\right) \\
&= \lim_{t\to 0} \frac{1}{t}\left(\int_0^{\tau(y)+t} q(\psi_s(y))\,ds - \int_0^{\tau(y)} q(\psi_s(y))\,ds\right) \\
&\quad - \lim_{t\to 0} \frac{1}{t}\int_0^t q(\psi_s(y))\,ds \\
&= q(\psi_{\tau(y)}(y)) - q(y) \\
&= q(T(y)) - q(y).
\end{aligned}$$

Therefore, I_g is T-cohomologous to I_h in Y.

Now we assume that I_g is T-cohomologous to I_h in Y. For each $x \in Y$, we have

$$\tau(\psi_t(x)) = \tau(x) - t$$

for any sufficiently small $t > 0$ (depending on x). Thus, $T(\psi_t(x)) = T(x)$ and it follows from (2.4) with $y = \psi_t(x)$ that

$$I_g(\psi_t(x)) - I_h(\psi_t(x)) = q(T(x)) - q(\psi_t(x)).$$

Since

$$\lim_{t\to 0^+} \frac{I_g(\psi_t(x)) - I_g(x)}{t} = \lim_{t\to 0} -\frac{1}{t}\int_0^t g(\psi_s(x))\,ds = -g(x),$$

we obtain

$$\begin{aligned} g(x) - h(x) &= \lim_{t\to 0^+}\left(-\frac{I_g(\psi_t(x)) - I_h(\psi_t(x))}{t} + \frac{I_g(x) - I_h(x)}{t}\right) \\ &= \lim_{t\to 0^+}\left(-\frac{q(T(x)) - q(\psi_t(x))}{t} + \frac{q(T(x)) - q(x)}{t}\right) \\ &= \lim_{t\to 0^+} \frac{q(\psi_t(x)) - q(x)}{t}. \end{aligned} \tag{2.5}$$

Similarly, we have

$$\tau(\psi_{-t}(x)) = \begin{cases} \tau(x) + t & \text{if } x \notin X \times \{0\}, \\ t & \text{if } x \in X \times \{0\} \end{cases}$$

for any sufficiently small $t > 0$ (depending on x). Now we consider two cases. For $x \notin X \times \{0\}$ we have $T(\psi_{-t}(x)) = T(x)$ and one can proceed in a similar manner to show that

$$g(x) - h(x) = \lim_{t\to 0^-} \frac{q(\psi_t(x)) - q(x)}{t}. \tag{2.6}$$

On the other hand, for $x \in X \times \{0\}$ we have $T(\psi_{-t}(x)) = x$ and it follows from (2.4) with $y = \psi_{-t}(x)$ that

$$I_g(\psi_{-t}(x)) - I_h(\psi_{-t}(x)) = q(x) - q(\psi_{-t}(x)).$$

Since

$$\lim_{t\to 0^+} \frac{I_g(\psi_{-t}(x))}{t} = \lim_{t\to 0^+} \frac{1}{t}\int_{-t}^0 g(\psi_s(x))\,ds = g(x),$$

we obtain

$$\begin{aligned} g(x) - h(x) &= \lim_{t\to 0^-} \frac{I_g(\psi_t(x)) - I_h(\psi_t(x))}{-t} \\ &= \lim_{t\to 0^-} \frac{q(x) - q(\psi_t(x))}{-t}. \end{aligned} \tag{2.7}$$

By (2.5), (2.6) and (2.7), for each $x \in Y$ we have

$$g(x) - h(x) = \lim_{t\to 0} \frac{q(\psi_t(x)) - q(x)}{t}$$

Therefore, g is Ψ-cohomologous to h in Y.

It remains to verify that Property 3 implies Property 2 (clearly, Property 2 implies Property 3). Let us assume that Property 3 holds for some function $q\colon X \times \{0\} \to \mathbb{R}$. We extend q to a function $q\colon Y \to \mathbb{R}$ by

$$q(\psi_t(y)) = q(y) - \int_0^t \big[g(\psi_s(y)) - h(\psi_s(y))\big]\, ds$$

for every $y = (x, 0)$ and $t \in [0, \tau(x))$. For each $t \in [0, \tau(x))$, we have $T(\psi_t(y)) = T(y)$ and by (2.3) we obtain

$$\begin{aligned} q(T(\psi_t(y))) - q(\psi_t(y)) &= q(T(y)) - q(\psi_t) \\ &= \int_t^{\tau(y)} \big[g(\psi_s(y)) - h(\psi_s(y))\big]\, ds \\ &= I_g(\psi_t(y)) - I_h(\psi_t(y)), \end{aligned}$$

which yields Property 2. This completes the proof of the theorem. □

By Theorem 2.1, each cohomology class for the dynamics in the base X induces a cohomology class for the suspension flow in the whole space Y, and all cohomology classes in Y appear in this way.

One can show that cohomologous functions have the same Birkhoff averages. These averages are one of the main elements of ergodic theory and multifractal analysis.

Theorem 2.2 *Let Ψ be a flow in Y and let $g, h\colon Y \to \mathbb{R}$ be continuous functions. If g and h are Ψ-cohomologous, then*

$$\liminf_{t\to\infty} \frac{1}{t} \int_0^t g(\psi_s(x))\, ds = \liminf_{t\to\infty} \frac{1}{t} \int_0^t h(\psi_s(x))\, ds \tag{2.8}$$

and

$$\limsup_{t\to\infty} \frac{1}{t} \int_0^t g(\psi_s(x))\, ds = \limsup_{t\to\infty} \frac{1}{t} \int_0^t h(\psi_s(x))\, ds \tag{2.9}$$

for every $x \in Y$.

Proof By (2.2), we have

$$\begin{aligned} \frac{1}{t}\int_0^t g(\psi_s(x))\, ds - \frac{1}{t}\int_0^t h(\psi_s(x))\, ds &= \frac{1}{t}\int_0^t \frac{d}{dt} q(\psi_t(\psi_s(x)))|_{t=0}\, ds \\ &= \frac{1}{t}\int_0^t \frac{d}{ds} q(\psi_s(x))\, ds \\ &= \frac{q(\psi_t(x)) - q(x)}{t}. \end{aligned} \tag{2.10}$$

The identities in (2.8) and (2.9) now follow readily from (2.10). □

It is also of interest to describe the convergence and divergence of the Birkhoff averages of the flow Ψ in terms of the Birkhoff averages of the map T in the base.

Theorem 2.3 *Let Ψ be a suspension flow over $T: X \to X$ with height function τ and let $g: Y \to \mathbb{R}$ be a continuous function. For each $x \in X$ and $s \in [0, \tau(x)]$, we have*

$$\liminf_{t\to\infty} \frac{1}{t} \int_0^t g(\psi_r(x,s))\,dr = \liminf_{m\to\infty} \frac{\sum_{i=0}^m I_g(T^i(x))}{\sum_{i=0}^m \tau(T^i(x))} \tag{2.11}$$

and

$$\limsup_{t\to\infty} \frac{1}{t} \int_0^t g(\psi_r(x,s))\,dr = \limsup_{m\to\infty} \frac{\sum_{i=0}^m I_g(T^i(x))}{\sum_{i=0}^m \tau(T^i(x))}. \tag{2.12}$$

Proof Given $m \in \mathbb{N}$, we define a function $\tau_m: Y \to \mathbb{R}$ by

$$\tau_m(x) = \sum_{i=0}^{m-1} \tau(T^i(x)). \tag{2.13}$$

For each $x \in Y$ and $m \in \mathbb{N}$, we have

$$\begin{aligned}\int_0^{\tau_m(x)} g(\psi_s(x))\,ds &= \sum_{i=0}^{m-1} \int_{\tau_i(x)}^{\tau_{i+1}(x)} g(\psi_s(x))\,ds \\ &= \sum_{i=0}^{m-1} \int_0^{\tau(T^i(x))} g(\psi_s(T^i(x)))\,ds \\ &= \sum_{i=0}^{m-1} I_g(T^i(x)).\end{aligned} \tag{2.14}$$

Now we observe that given $t > 0$, there exists a unique integer $m \in \mathbb{N}$ such that $\tau_m(x) \le t < \tau_{m+1}(x)$. We have $t = \tau_m(x) + \kappa$ for some $\kappa \in (\inf\tau, \sup\tau)$, and thus,

$$\frac{1}{t}\int_0^t g(\psi_s(x))\,ds = \frac{\int_0^{\tau_m(x)} g(\psi_s(x))\,ds + \int_{\tau_m(x)}^{\tau_m(x)+\kappa} g(\psi_s(x))\,ds}{\tau_m(x)+\kappa}.$$

Therefore,

$$\begin{aligned}
&\left|\frac{1}{t}\int_0^t g(\psi_s(x))\,ds - \frac{1}{\tau_m(x)}\int_0^{\tau_m(x)} g(\psi_s(x))\,ds\right| \\
&\le \left|\frac{1}{\tau_m(x)+\kappa} - \frac{1}{\tau_m(x)}\right| \int_0^{\tau_m(x)} |g(\psi_s(x))|\,ds + \frac{\kappa \sup|g|}{\tau_m(x)+\kappa} \\
&\le \frac{\kappa}{(\tau_m(x)+\kappa)\tau_m(x)} \cdot \tau_m(x)\sup|g| + \frac{\kappa \sup|g|}{\tau_m(x)+\kappa} \\
&\le \frac{2\sup\tau \sup|g|}{\tau_m(x)}.
\end{aligned}$$

Since τ is bounded (because it is a continuous function on the compact set X), letting $t \to \infty$, we have $m \to \infty$ and $\tau_m(x) \to \infty$. Hence, it follows from (2.14) that

$$\left|\frac{1}{t}\int_0^t g(\psi_s(x))\,ds - \frac{1}{\tau_m(x)}\sum_{i=0}^{m-1} I_g(T^i(x))\right| \to 0$$

when $t \to \infty$. This completes the proof of the theorem. □

We note that for each $x \in X$ the limits in (2.11) and (2.12) are independent of s.

More generally, one can consider a continuous map $T: X \to X$ that need not be a homeomorphism. More precisely, let T be a local homeomorphism in an open neighborhood of each point of the compact metric space X. The *suspension semiflow over T with height function τ* is the semiflow $\Psi = \{\psi_t\}_{t\in\mathbb{R}}$ in Y with the maps $\psi_t: Y \to Y$ defined by (2.1). One can readily extend Theorems 2.1 and 2.3 to suspension semiflows.

2.2 The Bowen–Walters Distance

In this section we describe a distance introduced by Bowen and Walters in [31] for suspension flows. We also establish several properties of this distance that are needed later on in the proofs.

Let $T: X \to X$ be a homeomorphism of a compact metric space (X, d_X) and let $\tau: X \to (0, \infty)$ be a Lipschitz function. We consider the suspension flow $\Psi = \{\psi_t\}_{t\in\mathbb{R}}$ in Y with the maps $\psi_t: Y \to Y$ given by (2.1). Without loss of generality, one can always assume that the diameter $\operatorname{diam} X$ of the space X is at most 1. When this is not the case, since X is compact, one can simply consider the new distance $d_X/\operatorname{diam} X$ in X.

We proceed with the construction of the Bowen–Walters distance. We first assume that $\tau = 1$. Given $x, y \in X$ and $t \in [0, 1]$, we define the length of the *horizontal segment* $[(x,t),(y,t)]$ (see Fig. 2.2) by

$$\rho_h((x,t),(y,t)) = (1-t)d_X(x,y) + t d_X(T(x),T(y)). \tag{2.15}$$

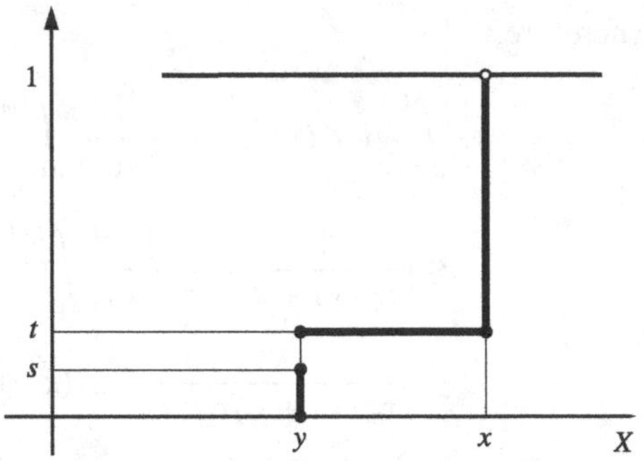

Fig. 2.2 Horizontal segment $[(x,t),(y,t)]$ and vertical segment $[(x,t),(y,s)]$

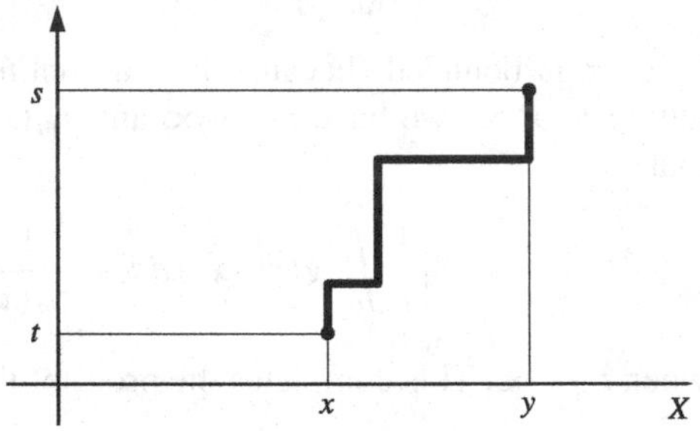

Fig. 2.3 A finite chain of horizontal and vertical segments between (x,t) and (y,s)

Clearly,

$$\rho_h((x,0),(y,0)) = d_X(x,y) \quad \text{and} \quad \rho_h((x,1),(y,1)) = d_X(T(x),T(y)).$$

Moreover, given points $(x,t),(y,s) \in Y$ in the same orbit, we define the length of the *vertical segment* $[(x,t),(y,s)]$ (see Fig. 2.2) by

$$\rho_v((x,t),(y,s)) = \inf\{|r| : \psi_r(x,t) = (y,s) \text{ and } r \in \mathbb{R}\}. \tag{2.16}$$

For the height function $\tau = 1$, the Bowen–Walters distance $d((x,t),(y,s))$ between two points $(x,t),(y,s) \in Y$ is defined as the infimum of the lengths of all paths between (x,t) and (y,s) that are composed of finitely many horizontal and vertical segments.

More precisely, for each $n \in \mathbb{N}$, we consider all finite chains

$$z_0 = (x,t),\ z_2,\ \ldots,\ z_{n-1},\ z_n = (y,s) \tag{2.17}$$

of points in Y such that for each $i = 0,\ldots,n-1$ the segment $[z_i, z_{i+1}]$ is either horizontal or vertical (see Fig. 2.3). The lengths of horizontal and vertical segments are defined respectively by (2.15) and (2.16). We remark that when the segment $[z_i, z_{i+1}]$ is simultaneously horizontal and vertical, since by hypothesis the space X has diameter at most 1, when computing the length of $[z_i, z_{i+1}]$ it is considered to be a horizontal segment. Finally, the length of the chain from z_0 to z_n in (2.17) is defined as the sum of the lengths of the segments $[z_i, z_{i+1}]$ for $i = 0,\ldots,n-1$.

Now we consider an arbitrary height function $\tau\colon X \to (0, \infty)$ and we introduce the Bowen–Walters distance d_Y in Y.

Definition 2.3 Given $(x, t), (y, s) \in Y$, we define

$$d_Y((x,t),(y,s)) = d\big((x, t/\tau(x)), (y, s/\tau(s))\big),$$

where d is the Bowen–Walters distance for the height function $\tau = 1$.

For an arbitrary function τ, a horizontal segment takes the form

$$w = [(x, t\tau(x)), (y, t\tau(y))],$$

and its length is given by

$$\ell_h(w) = (1-t)d_X(x,y) + td_X(T(x), T(y)).$$

Moreover, the length of a vertical segment $w = [(x,t),(x,s)]$ is now

$$\ell_v(w) = |t-s|/\tau(x),$$

for any sufficiently close t and s.

It is sometimes convenient to measure distances in another manner. Namely, given $(x, t), (y, s) \in Y$, let

$$d_\pi((x,t),(y,s)) = \min \left\{ \begin{array}{l} d_X(x,y) + |t-s|, \\ d_X(T(x), y) + \tau(x) - t + s, \\ d_X(x, T(y)) + \tau(y) - s + t \end{array} \right\}. \tag{2.18}$$

We note that d_π may not be a distance. Nevertheless, the following result relates d_π to the Bowen–Walters distance d_Y.

Proposition 2.1 *If T is an invertible Lipschitz map with Lipschitz inverse, then there exists a constant $c \geq 1$ such that*

$$c^{-1}d_\pi(p,q) \leq d_Y(p,q) \leq cd_\pi(p,q) \tag{2.19}$$

for every $p, q \in Y$.

Proof Let $(x,t),(y,s) \in Y$. One can easily verify that

$$L^{-1}|t-s| - L^2 d_X(x,y) \leq \left| \frac{t}{\tau(x)} - \frac{s}{\tau(y)} \right| \leq L|t-s| + L^2 d_X(x,y), \tag{2.20}$$

where $L \geq \max\{1/\min\tau, \sup\tau, 1\}$ is a Lipschitz constant simultaneously for T, T^{-1} and τ. Now we consider the chain formed by the points (x,t), $(y, t\tau(y)/\tau(x))$

and (y,s). It is composed of a horizontal segment and a vertical segment, and thus, using (2.20),

$$\begin{aligned} d_Y((x,t),(y,s)) &\le \ell_h\big((x,t),(y,t\tau(y)/\tau(x))\big) + \ell_v\big((y,t\tau(y)/\tau(x)),(y,s)\big) \\ &\le \left(1-\frac{t}{\tau(x)}\right) d_X(x,y) + \frac{t}{\tau(x)} d_X(T(x),T(y)) \\ &\quad + \left|\frac{t}{\tau(x)} - \frac{s}{\tau(y)}\right| \\ &\le L d_X(x,y) + L|t-s| + L^2 d_X(x,y). \end{aligned} \tag{2.21}$$

Therefore,

$$d_Y((x,t),(y,s)) \le c\big[d_X(x,y) + |t-s|\big] \tag{2.22}$$

taking $c \ge L + L^2$. Similarly, considering the chain formed by the points (x,t), $(x,\tau(x)) = (T(x),0)$, $(y,0)$ and (y,s), we obtain

$$\begin{aligned} d_Y((x,t),(y,s)) &\le \frac{\tau(x)-t}{\tau(x)} + d_X(T(x),y) + \frac{s}{\tau(y)} \\ &\le L\big[d_X(T(x),y) + \tau(x) - t + s\big]. \end{aligned} \tag{2.23}$$

By (2.22), (2.23) and the symmetry of d_Y, we conclude that

$$d_Y((x,t),(y,s)) \le c d_\pi((x,t),(y,s))$$

taking $c \ge L + L^2$.

For the other inequality in (2.19), consider a chain $z_0, \dots, z_n$ between (x,t) and (y,s) not intersecting the roof $\{(x,\tau(x)) : x \in X\}$ of Y. Let

$$\ell_H = \sum_{i\in H} \ell_h(z_i, z_{i+1}) \quad \text{and} \quad \ell_V = \sum_{i \in V} \ell_v(z_i, z_{i+1}),$$

where H is the set of all is such that $[z_i, z_{i+1}]$ is a horizontal segment and $V = \{0,\dots,n-1\} \setminus H$. We also write $z_i = (x_i, r_i) \in Y$. Since the chain does not intersect the roof, we obtain

$$\begin{aligned} \ell_H &= \sum_{i\in H} (1-r_i) d_X(x_i,x_{i+1}) + r_i d_X(T(x_i),T(x_{i+1})) \\ &\ge L^{-1} \sum_{i \in H} (1-r_i) d_X(x_i,x_{i+1}) + r_i d_X(x_i,x_{i+1}) \\ &\ge L^{-1} d_X(x,y). \end{aligned} \tag{2.24}$$

On the other hand, by (2.20) we have

$$\ell_V \ge |t/\tau(x) - s/\tau(y)| \ge L^{-1}|t-s| - L^2 d_X(x,y). \tag{2.25}$$

It follows from (2.24) and (2.25) that

$$2L^4\ell(z_0,\dots,z_n)\ge(L^4+L)\ell_H+L\ell_V\ge d_X(x,y)+|t-s|, \tag{2.26}$$

where $\ell(z_0,\dots,z_n)$ is the length of the chain $z_0,\dots,z_n$.

It is easy to verify that for any chain of length ℓ there exists a chain with the same endpoints, and of length at most $L\ell$, such that at most one segment of this chain intersects the roof. We notice that if a chain intersects the roof at least twice in the same direction, then its length is at least 2, which is larger than the length of the chain used to establish (2.21). This implies that $Ld_Y((x,t),(y,s))$ is bounded from below by the infimum of the lengths of all chains between (x,t) and (y,s) intersecting the roof at most once. Now let $z_0,\dots,z_n$ be a chain intersecting the roof exactly once. Without loss of generality, one can assume that there exists a $j\in\{1,\dots,n\}$ such that $r_j=\tau(x_j)$, with $z_j=(x_j,r_j)$, and $[z_{j-1},z_j]$ is a vertical segment. If the point z_j is after z_{j-1} on the same orbit, then by (2.26) we obtain

$$2L^4\big[\ell(z_0,\dots,z_j)+\ell(z_j,\dots,z_n)\big]\ge d_X(x,x_j)+\tau(x)-t+d_X(T(x_j),y)+s.$$

Since

$$Ld(x,x_j)+d(T(x_j),y)\ge d(T(x),T(x_j))+d(T(x_j),y)\ge d(T(x),y),$$

we conclude that

$$2L^5\ell(z_1,\dots,z_n)\ge d_X(T(x),y)+\tau(x)-t+s. \tag{2.27}$$

Similarly, if the point z_j is before z_{j-1} on the same orbit, then

$$2L^5\ell(z_1,\dots,z_n)\ge d_X(x,T(y))+\tau(y)-s+t. \tag{2.28}$$

By (2.26), (2.27) and (2.28), we obtain

$$d_\pi((x,t),(y,s))\le cd_Y((x,t),(y,s))$$

provided that $c\ge 2L^6$. Since $2L^6\ge L+L^2$, taking $c=2L^6$ we obtain the inequalities in (2.19). □

2.3 Further Properties

In this section we establish some additional properties of suspension flows, related to the existence and uniqueness of equilibrium and Gibbs measures. We continue to assume that T is an invertible Lipschitz map with Lipschitz inverse. Given a continuous function $g\colon Y\to\mathbb{R}$, we consider the function $I_g\colon X\to\mathbb{R}$ given by (2.3).

Proposition 2.2 *If g is Hölder continuous, then I_g is Hölder continuous in X.*

Proof Take $x, y \in X$ with $\tau(x) \ge \tau(y)$. We have

$$\begin{aligned}|I_g(x) - I_g(y)| &= \left| \int_{\tau(y)}^{\tau(x)} g(\psi_s(x))\,ds + \int_0^{\tau(y)} [g(\psi_s(x)) - g(\psi_s(y))]\,ds \right| \\ &\le \sup|g| \cdot |\tau(x) - \tau(y)| + \sup\tau \cdot \sup_{s\in(0,\tau(y))} |g(\psi_s(x)) - g(\psi_s(y))| \\ &\le \sup|g| \cdot L d_X(x, y) + b \sup_{s\in(0,\tau(y))} d_Y((x, s), (y, s))^\alpha \end{aligned} \tag{2.29}$$

for some positive constants α and b. It follows from Proposition 2.1 and (2.29) (together with (2.18)) that

$$\begin{aligned}|I_g(x) - I_g(y)| &\le \sup|g| \cdot L d_X(x, y) + b\big(c d_\pi((x, s), (y, s))\big)^\alpha \\ &\le \big[\sup|g| \cdot L + bc^\alpha\big] d_X(x, y)^\alpha.\end{aligned}$$

This yields the desired result. □

Now we consider Bowen balls in X and Y, defined respectively by

$$B_X(x, m, \varepsilon) = \bigcap_{0\le n\le m} T^{-n} B_X(T^n(x), \varepsilon) \tag{2.30}$$

for each $x \in X$, $m \in \mathbb{N}$ and $\varepsilon > 0$, and

$$B_Y(y, \rho, \varepsilon) = \bigcap_{0\le t\le \rho} \psi_{-t} B_Y(\psi_t(y), \varepsilon) \tag{2.31}$$

for each $y \in Y$ and $\rho, \varepsilon > 0$.

Definition 2.4 The map T is said to have *bounded variation* if for each Hölder continuous function $g : X \to \mathbb{R}$ there exists a constant $D > 0$ such that

$$\left| \sum_{k=0}^{m-1} g(T^k(x)) - \sum_{k=0}^{m-1} g(T^k(y)) \right| \le D\varepsilon$$

for every $x \in X$, $m \in \mathbb{N}$, $\varepsilon > 0$ and $y \in B_X(x, m, \varepsilon)$.

The following result establishes a relation between the Bowen balls in (2.30) and (2.31). We recall the function τ_m given by (2.13).

Proposition 2.3 *If T has bounded variation, then there exists a $\kappa \ge 1$ such that*

$$B_Y\big((x, s), \tau_m(x), \varepsilon/\kappa\big) \subset B_X(x, m, \varepsilon) \times (s - \varepsilon, s + \varepsilon) \subset B_Y\big((x, s), \tau_m(x), \kappa\varepsilon\big) \tag{2.32}$$

for every $x \in X$, $s \in [0, \tau(x)]$ and $m \in \mathbb{N}$, and any sufficiently small $\varepsilon > 0$.

Proof Take $\varepsilon \in (0, 1/(2c))$ with c as in Proposition 2.1. Moreover, take $(x,t) \in Y$ with $t \in (c\varepsilon, \tau(x) - c\varepsilon)$ and $(y,t) \in B_Y((x,s), \tau_m(x), \varepsilon)$.

If $m = 0$, then

$$d_\pi((x,t),(y,s)) \le c\varepsilon,$$

by Proposition 2.1. Since

$$\tau(x) - t + s \ge \tau(x) - t \ge c\varepsilon \quad \text{and} \quad \tau(y) - s + t \ge t \ge c\varepsilon,$$

we obtain

$$d_X(x,y) + |t - s| = d_\pi((x,t),(y,s)) \le c\varepsilon,$$

which implies that $d_X(x,y) \le c\varepsilon$ and $|t - s| \le c\varepsilon$. This establishes the first inclusion in (2.32) for $m = 0$.

Now take $n \in \{1, \dots, m\}$, and let $t_n = \tau_n(x) - t$ and $s_n = \tau_n(y) - s$. One can easily verify that

$$\psi_{t_n}(x,t) = (T^n(x), 0) \quad \text{and} \quad \psi_{s_n}(y,s) = (T^n(y), 0).$$

By Proposition 2.1, we obtain

$$\begin{aligned} d_X(T^n(x), T^n(y)) &\le c d_Y(\psi_{t_n}(x,t), \psi_{s_n}(y,s)) \\ &\le c d_Y(\psi_{t_n}(x,t), \psi_{t_n}(y,s)) + c d_Y(\psi_{t_n}(y,s), \psi_{s_n}(y,s)) \\ &\le c\varepsilon + c|t_n - s_n|. \end{aligned} \tag{2.33}$$

Moreover, by (2.18), we have

$$d_\pi(\psi_{t_n}(x,t), \psi_{t_n}(y,s)) \le c\varepsilon.$$

Thus, there exist $y_n \in X$ and $r_n \in (t_n - c\varepsilon, t_n + c\varepsilon)$ such that $\psi_{r_n}(y,s) = (y_n, 0)$. The sequence r_n is strictly increasing, because $t_{n+1} - t_n > 2c\varepsilon$. Hence $s_n \le r_n \le t_n + c\varepsilon$. By symmetry, we obtain $t_n \le s_n + c\varepsilon$, and hence $|t_n - s_n| \le c\varepsilon$. Finally, by (2.33), we conclude that

$$d_X(T^n(x), T^n(y)) \le c(1+c)\varepsilon.$$

Taking $\kappa \ge c(1+c)$, this establishes the first inclusion in (2.32).

Now let $y \in B_X(x, m, \varepsilon)$ and $s \in (t - \varepsilon, t + \varepsilon)$. Take $r \in (0, \tau_m(x))$ and $n \in \mathbb{N}$ such that

$$\tau_n(x) \le r + t < \tau_{n+1}(x).$$

We also write $r' = r + t - \tau_n(x) \geq 0$. Since T has bounded variation, it follows from Proposition 2.1 that

$$\begin{aligned}
d_Y(\psi_r(x,t), \psi_r(y,s)) &\leq d_Y\big((T^n(x), r'), (T^n(y), r')\big) + d_Y\big((T^n(y), r'), \psi_r(y,s)\big) \\
&\leq c d_\pi\big((T^n(x), r'), (T^n(y), r')\big) \\
&\quad + c d_\pi\big((T^n(y), r'), \psi_r(y,s)\big) \\
&\leq c d_X(T^n(x), T^n(y)) + c|r' + \tau_n(y) - r - s| \\
&\leq c d_X(T^n(x), T^n(y)) + c|t - s| + c|\tau_n(x) - \tau_n(y)| \\
&\leq c(2 + D)\varepsilon.
\end{aligned}$$

Taking $\kappa \geq c(2 + D)$, this establishes the second inclusion in (2.32). □

Chapter 3
Hyperbolic Flows

In this chapter we recall in a pragmatic manner all the necessary notions and results from hyperbolic dynamics, starting with the notion of a hyperbolic set for a flow. In particular, we consider the Markov systems constructed by Bowen and Ratner for a locally maximal hyperbolic set, and we describe how they can be used to associate a symbolic dynamics to the hyperbolic set. This allows one to see the restriction of any smooth flow to a hyperbolic set as a factor of a suspension flow over a topological Markov chain.

3.1 Basic Notions

Let $\Phi = \{\varphi_t\}_{t\in\mathbb{R}}$ be a C^1 flow in a smooth manifold M. This means that $\varphi_0 = \mathrm{id}$,

$$\varphi_t \circ \varphi_s = \varphi_{t+s} \quad \text{for} \quad t, s \in \mathbb{R},$$

and that the map $(t, x) \mapsto \varphi_t(x)$ is of class C^1. We first introduce the notion of a hyperbolic set.

Definition 3.1 A compact Φ-invariant set $\Lambda \subset M$ is said to be a *hyperbolic set* for Φ if there exists a splitting

$$T_\Lambda M = E^s \oplus E^u \oplus E^0,$$

(see Fig. 3.1) and constants $c > 0$ and $\lambda \in (0, 1)$ such that for each $x \in \Lambda$:

1. the vector $(d/dt)\varphi_t(x)|_{t=0}$ generates $E^0(x)$;
2. for each $t \in \mathbb{R}$ we have

$$d_x\varphi_t E^s(x) = E^s(\varphi_t(x)) \quad \text{and} \quad d_x\varphi_t E^u(x) = E^u(\varphi_t(x));$$

3. $\|d_x\varphi_t v\| \le c\lambda^t \|v\|$ for $v \in E^s(x)$ and $t > 0$;
4. $\|d_x\varphi_{-t} v\| \le c\lambda^t \|v\|$ for $v \in E^u(x)$ and $t > 0$.

L. Barreira, *Dimension Theory of Hyperbolic Flows*,
Springer Monographs in Mathematics, DOI 10.1007/978-3-319-00548-5_3,

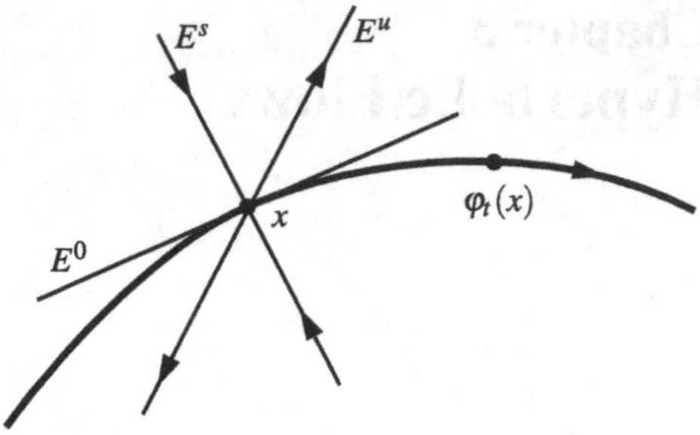

Fig. 3.1 The splitting of a hyperbolic set

For example, for any geodesic flow in a compact Riemannian manifold with strictly negative sectional curvature the whole unit tangent bundle is a hyperbolic set. Furthermore, time changes and small C^1 perturbations of a flow with a hyperbolic set also have a hyperbolic set.

Now let Λ be a hyperbolic set for Φ. For each $x \in \Lambda$ and any sufficiently small $\varepsilon > 0$, we consider the sets

$$A^s(x) = \{y \in B(x, \varepsilon) : d(\varphi_t(y), \varphi_t(x)) \searrow 0 \text{ when } t \to +\infty\}$$

and

$$A^u(x) = \{y \in B(x, \varepsilon) : d(\varphi_t(y), \varphi_t(x)) \searrow 0 \text{ when } t \to -\infty\}.$$

Let $V^s(x) \subset A^s(x)$ and $V^u(x) \subset A^u(x)$ be the largest connected components containing x. These are smooth manifolds, called respectively *(local) stable* and *unstable manifolds* (of size ε) at the point x. Moreover:

1.
$$T_x V^s(x) = E^s(x) \quad \text{and} \quad T_x V^u(x) = E^u(x);$$
2. for each $t > 0$ we have
$$\varphi_t(V^s(x)) \subset V^s(\varphi_t(x)) \quad \text{and} \quad \varphi_{-t}(V^u(x)) \subset V^u(\varphi_{-t}(x));$$
3. there exist $\kappa > 0$ and $\mu \in (0, 1)$ such that for each $t > 0$ we have
$$d(\varphi_t(y), \varphi_t(x)) \leq \kappa\mu^t d(y, x) \quad \text{for} \quad y \in V^s(x), \tag{3.1}$$
and
$$d(\varphi_{-t}(y), \varphi_{-t}(x)) \leq \kappa\mu^t d(y, x) \quad \text{for} \quad y \in V^u(x).$$

We also introduce the notion of a locally maximal hyperbolic set.

Definition 3.2 A set Λ is said to be *locally maximal* (with respect to a flow Φ) if there exists an open neighborhood U of Λ such that

$$\Lambda = \bigcap_{t \in \mathbb{R}} \varphi_t(U). \tag{3.2}$$

Now let Λ be a locally maximal hyperbolic set. For any sufficiently small $\varepsilon > 0$, there exists a $\delta > 0$ such that if $x, y \in \Lambda$ are at a distance $d(x, y) \le \delta$, then there exists a unique $t = t(x, y) \in [-\varepsilon, \varepsilon]$ for which the set

$$[x, y] = V^s(\varphi_t(x)) \cap V^u(y) \tag{3.3}$$

is a single point in Λ.

3.2 Markov Systems

In order to establish some of the results we need the notion of a Markov system and its associated symbolic dynamics. These were developed by Bowen [27] and Ratner [90].

Let Φ be a C^1 flow with a locally maximal hyperbolic set Λ. Consider an open smooth disk $D \subset M$ of dimension $\dim M - 1$ that is transverse to the flow Φ, and take $x \in D$. Let also $U(x)$ be an open neighborhood of x diffeomorphic to the product $D \times (-\varepsilon, \varepsilon)$. The projection $\pi_D \colon U(x) \to D$ defined by $\pi_D(\varphi_t(y)) = y$ is differentiable.

Definition 3.3 A closed set $R \subset \Lambda \cap D$ is said to be a *rectangle* if $R = \overline{\operatorname{int} R}$ (with the interior computed with respect to the induced topology on $\Lambda \cap D$) and $\pi_D([x, y]) \in R$ for $x, y \in R$.

Now we consider a collection of rectangles $R_1, \dots, R_k \subset \Lambda$ (each contained in some open disk transverse to the flow) such that

$$R_i \cap R_j = \partial R_i \cap \partial R_j \quad \text{for} \quad i \ne j.$$

Let $Z = \bigcup_{i=1}^k R_i$. We assume that there exists an $\varepsilon > 0$ such that:

1. $\Lambda = \bigcup_{t \in [0,\varepsilon]} \varphi_t(Z)$;
2. for each $i \ne j$ either

$$\varphi_t(R_i) \cap R_j = \varnothing \quad \text{for every} \quad t \in [0, \varepsilon],$$

   or

$$\varphi_t(R_j) \cap R_i = \varnothing \quad \text{for every} \quad t \in [0, \varepsilon].$$

We define the *transfer function* $\tau \colon \Lambda \to [0, \infty)$ by

$$\tau(x) = \min\{t > 0 : \varphi_t(x) \in Z\}, \tag{3.4}$$

and the *transfer map* $T \colon \Lambda \to Z$ by

$$T(x) = \varphi_{\tau(x)}(x). \tag{3.5}$$

The set Z is a Poincaré section for the flow Φ. One can easily verify that the restriction of the map T to Z is invertible. We also have $T^n(x) = \varphi_{\tau_n(x)}(x)$, where

$$\tau_n(x) = \sum_{i=0}^{n-1} \tau(T^i(x)). \tag{3.6}$$

Now we introduce the notion of a Markov system.

Definition 3.4 The collection of rectangles $R_1, \dots, R_k$ is said to be a *Markov system* for Φ on Λ if

$$T\big(\operatorname{int}(V^s(x) \cap R_i)\big) \subset \operatorname{int}\big(V^s(T(x)) \cap R_j\big)$$

and

$$T^{-1}\big(\operatorname{int}(V^u(T(x)) \cap R_j)\big) \subset \operatorname{int}\big(V^u(x) \cap R_i\big)$$

for every $x \in \operatorname{int} T(R_i) \cap \operatorname{int} R_j$.

It follows from work of Bowen and Ratner that any locally maximal hyperbolic set Λ has Markov systems of arbitrary small diameter (see [27, 90]). Furthermore, the map τ is Hölder continuous on each domain of continuity, and

$$0 < \inf\{\tau(x) : x \in Z\} \le \sup\{\tau(x) : x \in \Lambda\} < \infty. \tag{3.7}$$

3.3 Symbolic Dynamics

In this section we describe how a Markov system for a hyperbolic set gives rise to a symbolic dynamics.

Given a Markov system $R_1, \dots, R_k$ for a flow Φ on a locally maximal hyperbolic set Λ, we consider the $k \times k$ matrix A with entries

$$a_{ij} = \begin{cases} 1 & \text{if } \operatorname{int} T(R_i) \cap \operatorname{int} R_j \ne \varnothing, \\ 0 & \text{otherwise}, \end{cases} \tag{3.8}$$

where T is the transfer map in (3.5). We also consider the set $\Sigma_A \subset \{1, \dots, k\}^{\mathbb{Z}}$ given by

$$\Sigma_A = \big\{(\cdots i_{-1} i_0 i_1 \cdots) : a_{i_n i_{n+1}} = 1 \text{ for } n \in \mathbb{Z}\big\},$$

and the shift map $\sigma \colon \Sigma_A \to \Sigma_A$ defined by $\sigma(\cdots i_0 \cdots) = (\cdots j_0 \cdots)$, where $j_n = i_{n+1}$ for each $n \in \mathbb{Z}$.

Definition 3.5 The map $\sigma | \Sigma_A$ is said to be a *(two-sided) topological Markov chain* with *transition matrix* A.

We define a *coding map* $\pi\colon \Sigma_A \to \bigcup_{i=1}^k R_i$ for the hyperbolic set by

$$\pi(\cdots i_0 \cdots) = \bigcap_{j\in\mathbb{Z}} \overline{(T|Z)^{-j}(\operatorname{int} R_{i_j})}.$$

One can easily verify that

$$\pi \circ \sigma = T \circ \pi. \tag{3.9}$$

Given $\beta > 1$, we equip Σ_A with the distance d given by

$$d\big((\cdots i_{-1} i_0 i_1 \cdots), (\cdots j_{-1} j_0 j_1 \cdots)\big) = \sum_{n=-\infty}^{\infty} \beta^{-|n|} |i_n - j_n|. \tag{3.10}$$

As observed in [27], it is always possible to choose the constant β so that the function $\tau \circ \pi\colon \Sigma_A \to [0, \infty)$ is Lipschitz.

By (3.9), the restriction of a smooth flow to a locally maximal hyperbolic set is a factor of a suspension flow over a topological Markov chain. Namely, to each Markov system one can associate the suspension flow $\Psi = \{\psi_t\}_{t\in\mathbb{R}}$ over $\sigma|\Sigma_A$ with (Lipschitz) height function $\tau \circ \pi$. We extend π to a finite-to-one onto map $\pi\colon Y \to \Lambda$ by

$$\pi(x, s) = (\varphi_s \circ \pi)(x) \tag{3.11}$$

for $(x, s) \in Y$. Then

$$\pi \circ \psi_t = \varphi_t \circ \pi \tag{3.12}$$

for every $t \in \mathbb{R}$.

We denote by Σ_A^+ the set of (one-sided) sequences $(i_0 i_1 \cdots)$ such that

$$(i_0 i_1 \cdots) = (j_0 j_1 \cdots) \quad \text{for some} \quad (\cdots j_{-1} j_0 j_1 \cdots) \in \Sigma_A,$$

and by Σ_A^- the set of (one-sided) sequences $(\cdots i_{-1} i_0)$ such that

$$(\cdots i_{-1} i_0) = (\cdots j_{-1} j_0) \quad \text{for some} \quad (\cdots j_{-1} j_0 j_1 \cdots) \in \Sigma_A.$$

The set Σ_A^- can be identified with $\Sigma_{A^*}^+$, where A^* is the transpose of A, by the map

$$\Sigma_A^- \ni (\cdots i_{-1} i_0) \mapsto (i_0 i_{-1} \cdots) \in \Sigma_{A^*}^+.$$

We also consider the shift maps $\sigma_+\colon \Sigma_A^+ \to \Sigma_A^+$ and $\sigma_-\colon \Sigma_A^- \to \Sigma_A^-$ defined by

$$\sigma_+(i_0 i_1 \cdots) = (i_1 i_2 \cdots) \quad \text{and} \quad \sigma_-(\cdots i_{-1} i_0) = (\cdots i_{-2} i_{-1}).$$

Now we describe how distinct points in a stable or unstable manifold can be characterized in terms of the symbolic dynamics. Given $x \in \Lambda$, take $\omega \in \Sigma_A$ such that $\pi(\omega) = x$. Let $R(x)$ be a rectangle of the Markov system that contains x. For each $\omega' \in \Sigma_A$, we have

$$\pi(\omega') \in V^u(x) \cap R(x) \quad \text{whenever} \quad \pi_-(\omega') = \pi_-(\omega),$$

and

$$\pi(\omega') \in V^s(x) \cap R(x) \quad \text{whenever} \quad \pi_+(\omega') = \pi_+(\omega),$$

where $\pi_+ \colon \Sigma_A \to \Sigma_A^+$ and $\pi_- \colon \Sigma_A \to \Sigma_A^-$ are the projections defined by

$$\pi_+(\cdots i_{-1} i_0 i_1 \cdots) = (i_0 i_1 \cdots) \tag{3.13}$$

and

$$\pi_-(\cdots i_{-1} i_0 i_1 \cdots) = (\cdots i_{-1} i_0). \tag{3.14}$$

Therefore, writing $\omega = (\cdots i_{-1} i_0 i_1 \cdots)$, the set $V^u(x) \cap R(x)$ can be identified with the cylinder set

$$C_{i_0}^+ = \{(j_0 j_1 \cdots) \in \Sigma_A^+ : j_0 = i_0\} \subset \Sigma_A^+, \tag{3.15}$$

and the set $V^s(x) \cap R(x)$ can be identified with the cylinder set

$$C_{i_0}^- = \{(\cdots j_{-1} j_0) \in \Sigma_A^- : j_0 = i_0\} \subset \Sigma_A^-. \tag{3.16}$$

Chapter 4
Pressure and Dimension

In this chapter we recall in a pragmatic manner all the necessary notions and results from the thermodynamic formalism and dimension theory. In particular, we introduce the notions of topological pressure, BS-dimension, Hausdorff dimension, lower and upper box dimensions and pointwise dimension. We emphasize that we consider the general case of the topological pressure for noncompact sets, which is crucial in multifractal analysis.

4.1 Topological Pressure and Entropy

This section is dedicated to the notion of topological pressure and some of its basic properties, including the variational principle and its regularity properties. We refer the reader to [30, 63, 80, 92, 106] for details and proofs.

4.1.1 Basic Notions

We first introduce the notions of topological pressure and topological entropy.

Let $\Phi = \{\varphi_t\}_{t\in\mathbb{R}}$ be a continuous flow in a compact metric space (X, d). Given $x \in X$, $t > 0$ and $\varepsilon > 0$, we consider the Bowen ball

$$B(x,t,\varepsilon) = \big\{y \in X : d(\varphi_s(y), \varphi_s(x)) < \varepsilon \text{ for } s \in [0,t]\big\}. \tag{4.1}$$

Now let $a\colon X \to \mathbb{R}$ be a continuous function and write

$$a(x,t,\varepsilon) = \sup\left\{\int_0^t a(\varphi_s(y))\,ds : y \in B(x,t,\varepsilon)\right\}. \tag{4.2}$$

For each set $Z \subset X$ and $\alpha \in \mathbb{R}$, we define

$$M(Z,a,\alpha,\varepsilon) = \lim_{T\to\infty} \inf_{\Gamma} \sum_{(x,t)\in\Gamma} \exp\big(a(x,t,\varepsilon) - \alpha t\big),$$

L. Barreira, *Dimension Theory of Hyperbolic Flows*,
Springer Monographs in Mathematics, DOI 10.1007/978-3-319-00548-5_4,

where the infimum is taken over all finite or countable sets $\Gamma = \{(x_i, t_i)\}_{i\in I}$ such that $x_i \in X$ and $t_i \geq T$ for $i \in I$, and $\bigcup_{i\in I} B(x_i, t_i, \varepsilon) \supset Z$. One can easily verify that the limit

$$P_{\Phi|Z}(a) = \lim_{\varepsilon\to 0} \inf\bigl\{\alpha \in \mathbb{R} : M(Z, a, \alpha, \varepsilon) = 0\bigr\}$$

exists.

Definition 4.1 The number $P_{\Phi|Z}(a)$ is called the *topological pressure of a on the set Z* (with respect to the flow Φ).

We note that Z need not be compact nor Φ-invariant. This observation is crucial in multifractal analysis since typically the sets under consideration are not compact. For simplicity of notation, we also write $P_\Phi(a) = P_{\Phi|X}(a)$. When the set Z is compact and Φ-invariant, the topological pressure is given by

$$\begin{aligned} P_{\Phi|Z}(u) &= \lim_{\varepsilon\to 0} \liminf_{t\to\infty} \frac{1}{t} \log \inf_{\Gamma} \sum_{x\in\Gamma} \exp(u(x, t, \varepsilon)) \\ &= \lim_{\varepsilon\to 0} \limsup_{t\to\infty} \frac{1}{t} \log \inf_{\Gamma} \sum_{x\in\Gamma} \exp(u(x, t, \varepsilon)), \end{aligned} \tag{4.3}$$

where the infimum is taken over all finite or countable sets $\Gamma = \{x_i\}_{i\in I} \subset X$ such that $\bigcup_{i\in I} B(x_i, t, \varepsilon) \supset Z$.

Definition 4.2 The number $h(\Phi|Z) = P_{\Phi|Z}(0)$ is called the *topological entropy* of Φ on the set Z.

When Z is compact and Φ-invariant, we recover the usual notion of topological entropy, that is,

$$h(\Phi|Z) = \lim_{\varepsilon\to 0} \liminf_{t\to\infty} \frac{\log N_Z(t, \varepsilon)}{t} = \lim_{\varepsilon\to 0} \limsup_{t\to\infty} \frac{\log N_Z(t, \varepsilon)}{t},$$

where $N_Z(t, \varepsilon)$ is the least number of sets $D(x) = B(x, t, \varepsilon)$ that are needed to cover Z.

Now let $\mathcal{M}$ be the set of all Φ-invariant probability measures on X. We recall that a measure μ in X is said to be *Φ-invariant* if $\mu(\varphi_t(A)) = \mu(A)$ for every set $A \subset X$ and $t \in \mathbb{R}$. When equipped with the weak* topology, the space $\mathcal{M}$ is compact and metrizable. We also recall that a measure μ on X is said to be *ergodic* if for any Φ-invariant set $A \subset X$ (that is, any set such that $\varphi_t(A) = A$ for every $t \in \mathbb{R}$) we have $\mu(A) = 0$ or $\mu(X \setminus A) = 0$.

Moreover, for each measure $\mu \in \mathcal{M}$ the limit

$$h_\mu(\Phi) = \lim_{\varepsilon\to 0} \inf\{h(Z, \varepsilon) : \mu(Z) = 1\} \tag{4.4}$$

exists, where

$$h(Z,\varepsilon) = \inf\{\alpha \in \mathbb{R} : M(Z,0,\alpha,\varepsilon) = 0\}.$$

Proposition 4.1 *If Φ is a continuous flow in a compact metric space and $\mu \in \mathcal{M}$ is an ergodic measure, then the number $h_\mu(\Phi)$ in (4.4) coincides with the entropy of Φ with respect to μ, that is, the entropy of the time-1 map φ_1 with respect to μ.*

The proof of Proposition 4.1 can be obtained from a simple modification of the proof of an analogous result established by Pesin in [81] in the case of discrete time.

4.1.2 Properties of the Pressure

In this section we recall some basic properties of the topological pressure, starting with the variational principle.

Proposition 4.2 (Variational Principle) *If Φ is a continuous flow in a compact metric space X and $a\colon X \to \mathbb{R}$ is a continuous function, then*

$$P_\Phi(a) = \sup\left\{h_\mu(\Phi) + \int_X a\,d\mu : \mu \in \mathcal{M}\right\}. \tag{4.5}$$

A measure $\mu \in \mathcal{M}$ is said to be an *equilibrium measure* for the function a (with respect to the flow Φ) if the supremum in (4.5) is attained at this measure, that is,

$$P_\Phi(a) = h_\mu(\Phi) + \int_X a\,d\mu.$$

We denote by $C(X)$ the space of all continuous functions $a\colon X \to \mathbb{R}$ equipped with the supremum norm, and by $D(X) \subset C(X)$ the family of continuous functions with a unique equilibrium measure. Given a finite set $K \subset C(X)$, we denote by span $K \subset C(X)$ the linear space generated by the functions in K.

Proposition 4.3 *If Φ is a continuous flow in a compact metric space X such that the map $\mu \mapsto h_\mu(\Phi)$ is upper semicontinuous, then:*

1. *any function $a \in C(X)$ has equilibrium measures, and $D(X)$ is dense in $C(X)$;*
2. *given $a, b \in C(X)$, the map $\mathbb{R} \ni t \mapsto P_\Phi(a+tb)$ is differentiable at $t = 0$ if and only if $a \in D(X)$, in which case the unique equilibrium measure μ_a for the function a is ergodic and satisfies*

$$\frac{d}{dt}P_\Phi(a+tb)|_{t=0} = \int_X b\,d\mu_a; \tag{4.6}$$

3. *if $\operatorname{span}\{a,b\} \subset D(X)$, then the function $t \mapsto P_\Phi(a+tb)$ is of class C^1.*

In order to give some examples of flows with an upper semicontinuous entropy, we introduce the notion of an expansive flow.

Definition 4.3 A flow Φ in a metric space X is said to be *expansive* if there exists an $\varepsilon > 0$ such that given $x, y \in X$ and a continuous function $s\colon \mathbb{R} \to \mathbb{R}$ with $s(0) = 0$ satisfying

$$d(\varphi_t(x), \varphi_{s(t)}(x)) < \varepsilon \quad \text{and} \quad d(\varphi_t(x), \varphi_{s(t)}(y)) < \varepsilon$$

for every $t \in \mathbb{R}$, we have $x = y$.

If Φ is an expansive flow, then the map $\mu \mapsto h_\mu(\Phi)$ is upper semicontinuous (see [106]). For example, if Λ is a hyperbolic set for a flow Φ, then $\Phi|\Lambda$ is expansive.

Also for flows with a hyperbolic set, Proposition 4.3 can be strengthened as follows. We first recall that Φ is said to be *topologically mixing* on Λ (or simply $\Phi|\Lambda$ is said to be *topologically mixing*) if for any nonempty open sets U and V intersecting Λ there exists an $s \in \mathbb{R}$ such that $\varphi_t(U) \cap V \cap \Lambda \neq \varnothing$ for every $t > s$.

Proposition 4.4 *If Φ is a C^1 flow with a locally maximal hyperbolic set Λ such that $\Phi|\Lambda$ is topologically mixing, then:*

1. *the map $\mu \mapsto h_\mu(\Phi)$ is upper semicontinuous;*
2. *each Hölder continuous function $a\colon \Lambda \to \mathbb{R}$ has a unique equilibrium measure;*
3. *given Hölder continuous functions $a, b\colon \Lambda \to \mathbb{R}$, the function $\mathbb{R} \ni t \mapsto P_\Phi(a + tb)$ is analytic and*

$$\frac{d^2}{dt^2} P_\Phi(a + tb) \geq 0 \quad \text{for} \quad t \in \mathbb{R}, \tag{4.7}$$

with equality if and only if b is Φ-cohomologous to a constant.

The first property in Proposition 4.4 follows from the fact that $\Phi|\Lambda$ is expansive. We recall that a function $a\colon \Lambda \to \mathbb{R}$ is said to be *Φ-cohomologous* to a function $b\colon \Lambda \to \mathbb{R}$ if there exists a bounded measurable function $q\colon \Lambda \to \mathbb{R}$ such that

$$a(x) - b(x) = \lim_{t \to 0} \frac{q(\varphi_t(x)) - q(x)}{t}$$

(see Definition 2.2), in which case $P_{\Phi|\Lambda}(a) = P_{\Phi|\Lambda}(b)$. In particular, if b is constant, then $P_{\Phi|\Lambda}(a) = h(\Phi|\Lambda) + b$.

4.1.3 The Case of Suspension Flows

In this section we consider the particular case of suspension flows and we explain how the topological pressure and the invariant measures for the flow are related to the corresponding notions in the base.

Let $\Psi = \{\psi_t\}_{t\in\mathbb{R}}$ be a suspension flow in Y, over a homeomorphism $T: X \to X$ of the compact metric space X, and let μ be a T-invariant probability measure on X. One can show that μ induces a Ψ-invariant probability measure ν on Y such that

$$\int_Y g\,d\nu = \int_X \int_0^{\tau(x)} g(x,s)\,ds\,d\mu(x) \Big/ \int_X \tau\,d\mu \tag{4.8}$$

for every continuous function $g: Y \to \mathbb{R}$. We notice that locally ν is the product of μ and Lebesgue measure. Moreover, any Ψ-invariant probability measure ν on Y is of this form for some T-invariant probability measure μ on X. We remark that identity (4.8) is equivalent to

$$\int_Y g\,d\nu = \int_X I_g\,d\mu \Big/ \int_X \tau\,d\mu, \tag{4.9}$$

where I_g is the function given by (2.3). Moreover, Abramov's entropy formula (see [1]) says that

$$h_\nu(\Psi) = \frac{h_\mu(T)}{\int_X \tau\,d\mu}. \tag{4.10}$$

Now let $g: Y \to \mathbb{R}$ be a continuous function. By (4.9) and (4.10), we have

$$h_\nu(\Psi) + \int_Y g\,d\nu = \frac{h_\mu(T) + \int_X I_g\,d\mu}{\int_X \tau\,d\mu} \tag{4.11}$$

for any T-invariant probability measure μ on X, where ν is the Ψ-invariant probability measure induced by μ on Y. Since $\tau > 0$, it follows from (4.11) that

$$P_\Psi(g) = 0 \quad \text{if and only if} \quad P_T(I_g) = 0,$$

where $P_T(I_g)$ is the topological pressure of I_g with respect to T. Therefore, when $P_\Psi(g) = 0$ the measure ν is an equilibrium measure for g (with respect to Ψ) if and only if μ is an equilibrium measure for $I_g|X$ (with respect to T). Moreover, since

$$\begin{aligned}\sup_\mu \frac{h_\mu(T) + \int_X (I_g - P_\Psi(g)\tau)\,d\mu}{\int_X \tau\,d\mu} &= \sup_\mu \frac{h_\mu(T) + \int_X I_g\,d\mu}{\int_X \tau\,d\mu} - P_\Psi(g)\\ &= \sup_\nu \left(h_\nu(\Phi) + \int_Y g\,d\nu\right) - P_\Psi(g) = 0,\end{aligned}$$

we have

$$P_T(I_g - P_\Psi(g)\tau) = 0.$$

4.2 BS-Dimension

In this section we recall a Carathéodory characteristic introduced by Barreira and Saussol in [12]. It is a generalization of the notion of topological entropy and is a version of a Carathéodory characteristic introduced by Barreira and Schmeling in [17] in the case of discrete time.

Let Φ be a continuous flow in a compact metric space X and let $u\colon X \to \mathbb{R}^+$ be a continuous function. For each set $Z \subset X$ and $\alpha \in \mathbb{R}$, we define

$$N(Z,u,\alpha,\varepsilon) = \lim_{T\to\infty} \inf_{\Gamma} \sum_{(x,t)\in\Gamma} \exp\bigl(-\alpha u(x,t,\varepsilon)\bigr),$$

where the infimum is taken over all finite or countable sets $\Gamma = \{(x_i,t_i)\}_{i\in I}$ such that $x_i \in X$ and $t_i \ge T$ for $i \in I$, and $\bigcup_{i\in I} B(x_i,t_i,\varepsilon) \supset Z$. Writing

$$\dim_{u,\varepsilon} Z = \inf\bigl\{\alpha \in \mathbb{R} : N(Z,u,\alpha,\varepsilon) = 0\bigr\},$$

one can show that the limit

$$\dim_u Z = \lim_{\varepsilon\to 0} \dim_{u,\varepsilon} Z$$

exists.

Definition 4.4 $\dim_u Z$ is called the *BS-dimension of Z (with respect to u)*.

When $u = 1$ the BS-dimension coincides with the topological entropy, that is, $\dim_u Z = h(\Phi|Z)$.

It follows easily from the definitions that the topological pressure and the BS-dimension are related as follows.

Proposition 4.5 *The unique root of the equation $P_{\Phi|Z}(-\alpha u) = 0$ is $\alpha = \dim_u Z$.*

Given a probability measure μ in X and $\varepsilon > 0$, let

$$\dim_{u,\varepsilon} \mu = \inf\bigl\{\dim_{u,\varepsilon} Z : \mu(Z) = 1\bigr\}.$$

One can easily verify that the limit

$$\dim_u \mu = \lim_{\varepsilon\to 0} \dim_{u,\varepsilon} \mu$$

exists.

Definition 4.5 $\dim_u \mu$ is called the *BS-dimension of μ (with respect to u)*.

For each ergodic measure $\mu \in \mathcal{M}$, we have

$$\dim_u \mu = \frac{h_\mu(\Phi)}{\int_X u\,d\mu}.$$

This identity can be obtained in a similar manner to that in the case of discrete time (see [3, Proposition 7.2.7]).

4.3 Hausdorff and Box Dimensions

In this section we review some notions and results from dimension theory, both for sets and measures. In particular, we introduce the notions of Hausdorff dimension, lower and upper box dimensions, and pointwise dimension. We refer the reader to [3, 41] for details and proofs.

4.3.1 Dimension of Sets

We define the *diameter* of a set $U \subset \mathbb{R}^m$ by

$$\operatorname{diam} U = \sup\{\|x - y\| : x, y \in U\},$$

and the *diameter* of a collection $\mathcal{U}$ of subsets of $\mathbb{R}^m$ by

$$\operatorname{diam} \mathcal{U} = \sup\{\operatorname{diam} U : U \in \mathcal{U}\}.$$

Given a set $Z \subset \mathbb{R}^m$ and $\alpha \in \mathbb{R}$, the *α-dimensional Hausdorff measure* of Z is defined by

$$m(Z, \alpha) = \lim_{\varepsilon \to 0} \inf_{\mathcal{U}} \sum_{U \in \mathcal{U}} (\operatorname{diam} U)^\alpha,$$

where the infimum is taken over all finite or countable collections $\mathcal{U}$ with $\operatorname{diam} \mathcal{U} \le \varepsilon$ such that $\bigcup_{U \in \mathcal{U}} U \supset Z$.

Definition 4.6 The *Hausdorff dimension* of $Z \subset \mathbb{R}^m$ is defined by

$$\dim_H Z = \inf\{\alpha \in \mathbb{R} : m(Z, \alpha) = 0\}.$$

The *lower* and *upper box dimensions* of $Z \subset \mathbb{R}^m$ are defined respectively by

$$\underline{\dim}_B Z = \liminf_{\varepsilon \to 0} \frac{\log N(Z, \varepsilon)}{-\log \varepsilon} \quad \text{and} \quad \overline{\dim}_B Z = \limsup_{\varepsilon \to 0} \frac{\log N(Z, \varepsilon)}{-\log \varepsilon},$$

where $N(Z, \varepsilon)$ is the least number of balls of radius ε that are needed to cover the set Z.

One can easily verify that

$$\dim_H Z \le \underline{\dim}_B Z \le \overline{\dim}_B Z. \tag{4.12}$$

4.3.2 Dimension of Measures

Now we introduce corresponding notions for measures and we relate them to the pointwise dimension. Let μ be a finite measure on a set $X \subset \mathbb{R}^m$.

Definition 4.7 The *Hausdorff dimension* and the *lower* and *upper box dimensions* of μ are defined respectively by

$$\dim_H \mu = \inf\{\dim_H Z : \mu(X \setminus Z) = 0\},$$

$$\underline{\dim}_B \mu = \lim_{\delta \to 0} \inf\{\underline{\dim}_B Z : \mu(Z) \geq \mu(X) - \delta\},$$

$$\overline{\dim}_B \mu = \lim_{\delta \to 0} \inf\{\overline{\dim}_B Z : \mu(Z) \geq \mu(X) - \delta\}.$$

One can easily verify that

$$\dim_H \mu = \lim_{\delta \to 0} \inf\{\dim_H Z : \mu(Z) \geq \mu(X) - \delta\} \tag{4.13}$$

(see [3]). Moreover, it follows from (4.12) and (4.13) that

$$\dim_H \mu \leq \underline{\dim}_B \mu \leq \overline{\dim}_B \mu. \tag{4.14}$$

The following quantities allow us to formulate a criterion for the coincidence of the three numbers in (4.14).

Definition 4.8 The *lower* and *upper pointwise dimensions* of the measure μ at the point $x \in X$ are defined respectively by

$$\underline{d}_\mu(x) = \liminf_{r \to 0} \frac{\log \mu(B(x,r))}{\log r} \quad \text{and} \quad \overline{d}_\mu(x) = \limsup_{r \to 0} \frac{\log \mu(B(x,r))}{\log r}.$$

The following criterion is due to Young [108].

Proposition 4.6 *If μ is a finite measure on X and there exists a constant $d \geq 0$ such that*

$$\underline{d}_\mu(x) = \overline{d}_\mu(x) = d$$

for μ-almost every $x \in X$, then

$$\dim_H \mu = \underline{\dim}_B \mu = \overline{\dim}_B \mu = d.$$

The following result expresses the Hausdorff dimension of a measure in terms of the lower pointwise dimension.

Proposition 4.7 *If μ is a finite measure on X, then the following properties hold:*

1. *if* $\underline{d}_\mu(x) \geq \alpha$ *for* μ*-almost every* $x \in X$, *then* $\dim_H \mu \geq \alpha$;
2. *if* $\underline{d}_\mu(x) \leq \alpha$ *for every* $x \in Z \subset X$, *then* $\dim_H Z \leq \alpha$;
3. *we have*

$$\dim_H \mu = \operatorname{ess\,sup}\{\underline{d}_\mu(x) : x \in X\}.$$

By Whitney's embedding theorem, Proposition 4.7 can be readily extended to measures on subsets of smooth manifolds.

We also want to describe how the Hausdorff dimension of an invariant measure is related to its ergodic decompositions. Let Φ be a continuous flow in a metric space M and let $X \subset M$ be a compact Φ-invariant set. We continue to denote by $\mathcal{M}$ the set of all Φ-invariant probability measures on X and we endow it with the weak* topology. Let also $\mathcal{M}_E \subset \mathcal{M}$ be the subset of all ergodic measures.

Definition 4.9 Given $\mu \in \mathcal{M}$, a probability measure τ in $\mathcal{M}$ (or, more precisely, in the Borel σ-algebra generated by the weak* topology) is said to be an *ergodic decomposition* of μ if $\tau(\mathcal{M}_E) = 1$ and

$$\int_X \varphi \, d\mu = \int_{\mathcal{M}} \left(\int_X \varphi \, d\nu \right) d\tau(\nu)$$

for any continuous function $\varphi \colon X \to \mathbb{R}$.

It is well known that any measure $\mu \in \mathcal{M}$ has ergodic decompositions. The following statement is a simple consequence of the definitions.

Proposition 4.8 *If* τ *is an ergodic decomposition of a measure* $\mu \in \mathcal{M}$, *then*

$$\dim_H \mu \geq \operatorname{ess\,sup}\{\dim_H \nu : \nu \in \mathcal{M}_E\}, \tag{4.15}$$

where the essential supremum is taken with respect to τ.

We emphasize that in general inequality (4.15) may be strict. For example, if Φ is a rational linear flow in the 2-torus $\mathbb{T}^2$, then the Lebesgue measure μ has Hausdorff dimension $\dim_H \mu = 2$ but clearly

$$\operatorname{ess\,sup}\{\dim_H \nu : \nu \in \mathcal{M}_E\} = 1.$$

Part II
Dimension Theory

This part is dedicated to the dimension theory of hyperbolic flows, both for invariant measures and invariant sets. In Chap. 5 we study the dimension of a locally maximal hyperbolic set for a conformal flow in terms of the topological pressure. The arguments use Markov systems. Chapter 6 is dedicated to the study of the pointwise dimension of an arbitrary invariant measure sitting on a locally maximal hyperbolic set for a conformal flow. The pointwise dimension is expressed in terms of the local entropy and the Lyapunov exponents. We also describe the Hausdorff dimension of a nonergodic measure in terms of an ergodic decomposition and we establish the existence of invariant measures of maximal dimension.

Chapter 5
Dimension of Hyperbolic Sets

This chapter is dedicated to the study of the dimension of a locally maximal hyperbolic set for a conformal flow. We first consider the dimensions along the stable and unstable manifolds and we compute them in terms of the topological pressure. We also show that the Hausdorff dimension and the lower and upper box dimensions of the hyperbolic set coincide and that they are obtained by adding the dimensions along the stable and unstable manifolds, plus the dimension along the flow. This is a consequence of the conformality of the flow. The proofs are based on the use of Markov systems.

5.1 Dimensions Along Stable and Unstable Manifolds

In this section we obtain formulas for the dimensions of a locally maximal hyperbolic set for a conformal flow along the stable and unstable manifolds. These are expressed in terms of the topological pressure.

Let $\Phi = \{\varphi_t\}_{t\in\mathbb{R}}$ be a C^1 flow and let Λ be a locally maximal hyperbolic set for Φ. We first introduce the notion of a conformal flow.

Definition 5.1 The flow Φ is said to be *conformal* on Λ (or simply $\Phi|\Lambda$ is said to be *conformal*) if the maps

$$d_x\varphi_t|E^s(x)\colon E^s(x)\to E^s(\varphi_t(x)) \quad \text{and} \quad d_x\varphi_t|E^u(x)\colon E^u(x)\to E^u(\varphi_t(x))$$

are multiples of isometries for every $x\in\Lambda$ and $t\in\mathbb{R}$.

This means that the flow contracts and expands equally in all directions. For example, if

$$\dim E^s(x) = \dim E^u(x) = 1 \tag{5.1}$$

for every $x\in\Lambda$, then the flow is conformal on Λ. Specific examples satisfying (5.1) are given by any geodesic flow on the unit tangent bundle of a compact surface M

L. Barreira, *Dimension Theory of Hyperbolic Flows*,
Springer Monographs in Mathematics, DOI 10.1007/978-3-319-00548-5_5,

with negative (sectional) curvature. Certainly, the dimension theory of this particular class of examples is trivial (because the whole unit tangent bundle is a hyperbolic set for the geodesic flow and thus, its dimension is simply $2 \dim M - 1$). On the other hand, the multifractal analysis of dimension spectra, also developed for conformal flows (see Chap. 8), is nontrivial even in this particular class of examples.

Now we assume that $\Phi|\Lambda$ is conformal and we consider the families of local stable and unstable manifolds $V^s(x)$ and $V^u(x)$ for $x \in \Lambda$. The following result of Pesin and Sadovskaya [82] expresses the dimensions of the sets $V^s(x) \cap \Lambda$ and $V^u(x) \cap \Lambda$ in terms of the topological pressure. We define functions $\zeta_s, \zeta_u \colon \Lambda \to \mathbb{R}$ by

$$\zeta_s(x) = \frac{\partial}{\partial t} \log\|d_x\varphi_t|E^s(x)\|\Big|_{t=0} = \lim_{t\to 0} \frac{1}{t} \log\|d_x\varphi_t|E^s(x)\| \tag{5.2}$$

and

$$\zeta_u(x) = \frac{\partial}{\partial t} \log\|d_x\varphi_t|E^u(x)\|\Big|_{t=0} = \lim_{t\to 0} \frac{1}{t} \log\|d_x\varphi_t|E^u(x)\|. \tag{5.3}$$

Since the flow Φ is of class C^1, these functions are well defined.

Theorem 5.1 *Let Φ be a $C^{1+\delta}$ flow with a locally maximal hyperbolic set Λ such that $\Phi|\Lambda$ is conformal and topologically mixing. Then*

$$\dim_H(V^s(x) \cap \Lambda) = \underline{\dim}_B(V^s(x) \cap \Lambda) = \overline{\dim}_B(V^s(x) \cap \Lambda) = t_s \tag{5.4}$$

and

$$\dim_H(V^u(x) \cap \Lambda) = \underline{\dim}_B(V^u(x) \cap \Lambda) = \overline{\dim}_B(V^u(x) \cap \Lambda) = t_u \tag{5.5}$$

for every $x \in \Lambda$, where t_s and t_u are the unique real numbers such that

$$P_{\Phi|\Lambda}(t_s\zeta_s) = P_{\Phi|\Lambda}(-t_u\zeta_u) = 0. \tag{5.6}$$

Proof The idea of the proof is to first compute the dimensions along the stable and unstable directions inside the elements of some Markov system. We start by verifying that the numbers t_s and t_u are well defined. It follows from (4.6) that

$$\frac{d}{dt} P_{\Phi|\Lambda}(t\zeta_s) = \int_\Lambda \zeta_s \, d\mu_s^t$$

and

$$\frac{d}{dt} P_{\Phi|\Lambda}(-t\zeta_u) = -\int_\Lambda \zeta_u \, d\mu_u^t,$$

where μ_s^t and μ_u^t are respectively the equilibrium measures for $t\zeta_s$ and $-t\zeta_u$. By Birkhoff's ergodic theorem, we have

$$\int_\Lambda \zeta_s \, d\mu_s^t = \int_\Lambda \lim_{\tau\to\infty} \frac{1}{\tau} \int_0^\tau \zeta_s(\varphi_v(x)) \, dv \, d\mu_s^t(x)$$

$$= \int_\Lambda \lim_{\tau\to\infty} \frac{1}{\tau} \log \|d_x\varphi_\tau | E^s(x)\| \, d\mu_s^t(x)$$
$$\le \log\lambda < 0, \tag{5.7}$$

and hence, the function $t \mapsto P_{\Phi|\Lambda}(t\zeta_s)$ is strictly decreasing. Moreover, $P_{\Phi|\Lambda}(0) = h(\Phi|\Lambda) \ge 0$. Thus, there exists a unique real number t_s such that $P_{\Phi|\Lambda}(t_s\zeta_s) = 0$ and $t_s \ge 0$. A similar argument shows that t_u is also uniquely defined and that $t_u \ge 0$.

Let $R_1, \ldots, R_k$ be a Markov system for Φ on Λ. We assume that the diameter of the rectangles R_i is small when compared to the sizes of the stable and unstable manifolds. We also consider the function τ in (3.4) and the map T in (3.5), where $Z = \bigcup_{i=1}^k R_i$.

We only establish the identities in (5.5). The argument for (5.4) is entirely analogous. For $i = 1, \ldots, k$, let

$$V_i = T(V^u(x) \cap \Lambda) \cap R_i \tag{5.8}$$

and $V = \bigcup_{i=1}^k V_i$. Let also S be the invertible map $T|Z\colon Z \to Z$. We define

$$R_{i_0\cdots i_n} = \bigcap_{j=0}^n S^{-j} R_{i_j} \quad \text{and} \quad V_{i_0\cdots i_n} = V \cap R_{i_0\cdots i_n} \tag{5.9}$$

for each $(\cdots i_0 \cdots) \in \Sigma_A$ and $n \in \mathbb{N}$, where A is the transition matrix obtained from the Markov system as in (3.8).

We first obtain an upper bound for the upper box dimension. Since $T^n V_{i_0\cdots i_n} \subset V_{i_n}$, if $\mathcal{U}$ is a cover of V_{i_n}, then $S^{-n}\mathcal{U}$ is a cover of $V_{i_0\cdots i_n}$. Therefore,

$$N(V_{i_0\cdots i_n}, r) \le N(V_{i_n}, r/\overline{\lambda}_{i_0\cdots i_n})$$

for $r > 0$, where

$$\overline{\lambda}_{i_0\cdots i_n} = \max\bigl\{\|d_x S^{-n} | E^u(x)\| : x \in R_{i_0\cdots i_n}\bigr\},$$

and hence,

$$N(V, r) \le \sum_{i_0\cdots i_n} N(V_{i_0\cdots i_n}, r) \le \sum_{i_0\cdots i_n} N(V, r/\overline{\lambda}_{i_0\cdots i_n}).$$

Now let us take $s > \overline{\dim}_B V$. Then there exists an $r_0 > 0$ such that $N(V, r) < r^{-s}$ for $r \in (0, r_0)$. Letting

$$c_n(s) = \sum_{i_0\cdots i_n} \overline{\lambda}_{i_0\cdots i_n}^s,$$

we obtain $N(V, r) \le r^{-s} c_n(s)$ for $r < \lambda_n r_0$, where

$$\lambda_n = \min_{i_0\cdots i_n} \overline{\lambda}_{i_0\cdots i_n}.$$

It follows by induction that

$$N(V,r) \le r^{-s} c_n(s)^m$$

for $m \in \mathbb{N}$ and $r < \lambda_n^m r_0$. Therefore,

$$\frac{\log N(V,r)}{-\log r} \le s + \frac{m \log c_n(s)}{-\log r} \le s + \frac{m \log c_n(s)}{-\log(\lambda_n^m r_0)},$$

and letting $r \to 0$ yields the inequality

$$\overline{\dim}_B V \le s + \limsup_{m\to\infty} \frac{m \log c_n(s)}{-\log(\lambda_n^m r_0)} = s - \frac{\log c_n(s)}{\log \lambda_n}.$$

Letting $s \searrow \overline{\dim}_B V$, it follows from this inequality that

$$c_n(\overline{\dim}_B V) \ge 1 \tag{5.10}$$

for any sufficiently large n (because then $\lambda_n < 1$). Now we observe that

$$c_n(s) = \sum_{i_0\cdots i_n} \exp \max_{x\in R_{i_0\cdots i_n}} \left(-s \int_0^{\tau_n(x)} \zeta_u(\varphi_v(x))\, dv \right),$$

with $\tau_n(x)$ as in (3.6). It follows from (4.3) and (5.10) that

$$P_{\Phi|\Lambda}(-s\zeta_u) = \lim_{n\to\infty} \frac{1}{n} \log c_n(s) \ge 0$$

(we note that in the present context the limits when $\varepsilon \to 0$ in (4.3) are not necessary). Since the function $s \mapsto P_{\Phi|\Lambda}(-s\zeta_u)$ is strictly decreasing and $P_{\Phi|\Lambda}(-t_u\zeta_u) = 0$, we conclude that

$$s \le t_u \quad \text{for} \quad s > \overline{\dim}_B V.$$

Finally, letting $s \to \overline{\dim}_B V$ yields the inequality $\overline{\dim}_B V \le t_u$.

Now we consider the Hausdorff dimension and we proceed by contradiction. Let us assume that $\dim_H V < t_u$ and take $s > 0$ such that

$$\dim_H V < s < t_u. \tag{5.11}$$

Then $m(V,s) = 0$, and since $\overline{V}$ is compact, given $\delta > 0$, there exists a finite open cover $\mathcal{U}$ of V such that

$$\sum_{U\in\mathcal{U}} (\operatorname{diam} U)^s < \delta^s. \tag{5.12}$$

For each $n \in \mathbb{N}$, take $\delta_n > 0$ such that

$$p_n(U) = \operatorname{card}\left\{(i_0\cdots i_n) : U \cap R_{i_0\cdots i_n} \ne \varnothing\right\} < k$$

whenever $\operatorname{diam} U < \delta_n$ (we recall that k is the number of elements of the Markov system). We note that $\delta_n \to 0$ when $n \to \infty$. It follows from (5.12) with $\delta = \delta_n$ that $\operatorname{diam} \mathfrak{U} < \delta_n$ and hence, $p_n(U) < k$ for every $U \in \mathfrak{U}$. Now let $N = n + m - 1$ for some $m \in \mathbb{N}$ such that all entries of the matrix A^m are positive (we recall that $\Phi|\Lambda$ is topologically mixing). For each $(i_0 \cdots) \in \Sigma_A^+$ and $n \in \mathbb{N}$, let $\mathfrak{U}_{i_0 \cdots i_N}$ be the cover of V composed of the sets $T^N(U)$ with $U \in \mathfrak{U}$ such that $U \cap R_{i_0 \cdots i_n} \neq \varnothing$. We have

$$\sum_{U \in \mathfrak{U}_{i_0 \cdots i_N}} (\operatorname{diam} U)^s \le \underline{\lambda}_{i_0 \cdots i_N}^{-s} \sum_{U \in \mathfrak{U},\, U \cap R_{i_0 \cdots i_n} \neq \varnothing} (\operatorname{diam} U)^s,$$

where

$$\underline{\lambda}_{i_0 \cdots i_n} = \min\{\|d_x S^{-n} | E^u(x)\| : x \in R_{i_0 \cdots i_n}\}.$$

Now let us assume that

$$\sum_{U \in \mathfrak{U}_{i_0 \cdots i_N}} (\operatorname{diam} U)^s \ge \delta_n^s$$

for every $(i_0 \cdots) \in \Sigma_A^+$ and $n \in \mathbb{N}$. We obtain

$$\begin{aligned} k\delta_n^s > k \sum_{U \in \mathfrak{U}} (\operatorname{diam} U)^s &\ge \sum_{U \in \mathfrak{U}} p_n(U)(\operatorname{diam} U)^s \\ &= \sum_{i_0 \cdots i_n} \sum_{U \in \mathfrak{U},\, U \cap R_{i_0 \cdots i_n} \neq \varnothing} (\operatorname{diam} U)^s \\ &\ge k^{-m+1} \sum_{i_0 \cdots i_N} \sum_{U \in \mathfrak{U},\, U \cap R_{i_0 \cdots i_n} \neq \varnothing} (\operatorname{diam} U)^s \\ &\ge k^{-m+1} \sum_{i_0 \cdots i_N} \Big(\underline{\lambda}_{i_0 \cdots i_N} \sum_{U \in \mathfrak{U}_{i_0 \cdots i_N}} (\operatorname{diam} U)^s \Big) \\ &\ge k^{-m+1} \delta_n^s \sum_{i_0 \cdots i_N} \underline{\lambda}_{i_0 \cdots i_N}^s \end{aligned}$$

and hence,

$$\sum_{i_0 \cdots i_N} \underline{\lambda}_{i_0 \cdots i_N}^s \le k^m. \tag{5.13}$$

Since the map

$$\psi(x) = \|d_x S^{-1} | E^u(x)\|$$

is Hölder continuous, for each $x, y \in R_{i_0 \cdots i_n}$ we have

$$\frac{\|d_x S^{-n} | E^u(x)\|}{\|d_y S^{-n} | E^u(y)\|} = \prod_{j=0}^{n-1} \frac{\psi(T^j(x))}{\psi(T^j(y))}$$

$$\leq \prod_{j=0}^{n-1}\left(1+\frac{|\psi(T^j(x))-\psi(T^j(y))|}{\inf\psi}\right)$$

$$\leq \prod_{j=0}^{n-1}\left(1+Kd\big(T^j(x),T^j(y)\big)^{\delta}\right)$$

$$\leq \prod_{j=0}^{n-1}\left(1+Kd\big(T^n(x),T^n(y)\big)^{\delta}\lambda^{\delta(n-j)}\right), \tag{5.14}$$

for some constants $K>0$ and $\lambda,\delta\in(0,1)$. Since $T^n(x),T^n(y)\in R_{i_0}$, it follows from (5.14) that

$$\frac{\|d_xS^{-n}|E^u(x)\|}{\|d_yS^{-n}|E^u(y)\|}\leq \prod_{j=0}^{n-1}(1+K'\lambda^{\delta(n-j)})$$

$$\leq \prod_{j=1}^{\infty}(1+K'\lambda^{\delta j})<\infty \tag{5.15}$$

for some constant $K'>0$. Hence, there exists an $L>0$ such that

$$\overline{\lambda}_{i_0\cdots i_n}\leq L\underline{\lambda}_{i_0\cdots i_n}$$

for every $(i_0\cdots)\in\Sigma_A^+$ and $n\in\mathbb{N}$. By (5.13), we obtain

$$P_{\Phi|\Lambda}(-s\zeta_u)=\lim_{N\to\infty}\frac{1}{N}\sum_{i_0\cdots i_N}\overline{\lambda}^s_{i_0\cdots i_n}$$

$$\leq \lim_{N\to\infty}\frac{1}{N}\sum_{i_0\cdots i_N}\underline{\lambda}^s_{i_0\cdots i_n}\leq 0.$$

Since the function $s\mapsto P_{\Phi|\Lambda}(-s\zeta_u)$ is strictly decreasing and $P_{\Phi|\Lambda}(-t_u\zeta_u)=0$, this contradicts (5.11). Therefore,

$$\sum_{U\in\mathfrak{U}_{i_0\cdots i_N}}(\operatorname{diam}U)^s<\delta_n^s \tag{5.16}$$

for some sequence $i_0\cdots i_N$ and any sufficiently large n (recall that $N=n+m-1$). Now we restart the process using the cover $\mathcal{V}_1=\mathfrak{U}_{i_0\cdots i_N}$ to find inductively finite covers $\mathcal{V}_l$ of V for each $l\in\mathbb{N}$. By (5.16), we have $\operatorname{diam}\mathcal{V}_l<\delta_n$ and hence, $p_n(U)<k$ for every $U\in\mathcal{V}_l$. This implies that $\operatorname{card}\mathcal{V}_{l+1}<\operatorname{card}\mathcal{V}_l$ and thus, $\operatorname{card}\mathcal{V}_l=1$ for some $l=l(n)$. Writing $\mathcal{V}_{l(n)}=\{U_n\}$, we obtain

$$\operatorname{diam}V\leq\operatorname{diam}U_n<\delta_n\to 0$$

when $n \to \infty$, which is impossible. This contradiction shows that $\dim_H V \geq t_u$.

We have shown that

$$\dim_H V = \underline{\dim}_B V = \overline{\dim}_B V = t_u. \tag{5.17}$$

Now we observe that the map $F = T|(V^u(x) \cap \Lambda)$ and its inverse are Lipschitz in each domain of continuity (we recall that each rectangle is contained in a smooth disk, which ensures that the restriction $\tau|(V^u(x) \cap \Lambda)$ is Lipschitz in each domain of continuity). In particular, F preserves the Hausdorff and box dimensions, and hence, the identities in (5.5) follow readily from (5.17). □

We emphasize that the dimensions of the sets $V^s(x) \cap \Lambda$ and $V^u(x) \cap \Lambda$ are independent of the point x (see (5.4) and (5.5)). Our proof of Theorem 5.1 is based on corresponding arguments of Barreira [2] in the case of discrete time (see [3] for details and further references). The original argument of Pesin and Sadovskaya in [82] uses Moran covers instead.

More generally, if $\Phi|\Lambda$ is not conformal but the maps

$$d_x\varphi_t|E^u(x)\colon E^u(x) \to E^u(\varphi_t(x))$$

are multiples of isometries for every $x \in \Lambda$ and $t \in \mathbb{R}$, then the identities in (5.5) still hold (without modifications in the proof of Theorem 5.1). A similar observation holds for the dimensions along the stable manifolds.

Now let Φ be a $C^{1+\delta}$ flow with a locally maximal hyperbolic set Λ such that $\Phi|\Lambda$ is topologically mixing but not necessarily conformal. It follows from the proof of Theorem 5.1 that if $R_1, \ldots, R_k$ is a Markov system for Φ on Λ, then

$$\dim_H(V^u(x) \cap \Lambda) \geq \underline{r} \quad \text{and} \quad \overline{\dim}_B(V^u(x) \cap \Lambda) \leq \overline{r} \tag{5.18}$$

for every $x \in \Lambda$, where $\underline{r}$ and $\overline{r}$ are the unique real numbers such that

$$\lim_{n\to\infty} \frac{1}{n} \log \sum_{i_0\cdots i_n} \exp \min_{x\in R_{i_0\cdots i_n}} \left(-\underline{r} \int_0^{\tau_n(x)} \zeta_u(\varphi_v(x))\, dv \right) = 0$$

and

$$\lim_{n\to\infty} \frac{1}{n} \log \sum_{i_0\cdots i_n} \exp \max_{x\in R_{i_0\cdots i_n}} \left(-\overline{r} \int_0^{\tau_n(x)} \zeta_u(\varphi_v(x))\, dv \right) = 0, \tag{5.19}$$

with the sets $R_{i_0\cdots i_n}$ as in (5.9). We note that the limit in (5.19) coincides with $P_{\Phi|\Lambda}(-\overline{r}\zeta_u)$, and hence, $\overline{r} = t_u$. Similar observations hold for the dimensions along the stable manifolds.

5.2 Formula for the Dimension

In this section we establish a formula for the dimension of a locally maximal hyperbolic set for a conformal flow. Due to the conformality, the dimension is obtained by adding the dimensions along the stable and unstable manifolds, plus the dimension along the flow.

Theorem 5.2 ([82]) *If Φ is a $C^{1+\delta}$ flow with a locally maximal hyperbolic set Λ such that $\Phi|\Lambda$ is conformal and topologically mixing, then*

$$\dim_H \Lambda = \underline{\dim}_B \Lambda = \overline{\dim}_B \Lambda = t_s + t_u + 1. \tag{5.20}$$

Proof Again, the idea of the proof is to reduce the problem to a Markov system. We then use the conformality of $\Phi|\Lambda$ to show that the dimensions in Theorem 5.1 can be added.

Let $R_1, \ldots, R_k$ be a Markov system for Φ on Λ. Each rectangle R_i is contained in a smooth disk D_i and has the product structure

$$\{x, y\} = \pi_i([x, y]),$$

where π_i is the projection onto D_i, with $[x, y]$ as in (3.2). In general, the map $(x, y) \mapsto \{x, y\}$ is only Hölder continuous and has a Hölder continuous inverse. However, since Φ is conformal on Λ, it follows from results of Hasselblatt in [53] that the distributions $x \mapsto E^s(x) \oplus E^0(x)$ and $x \mapsto E^u(x) \oplus E^0(x)$ are Lipschitz. This implies that

$$R_i \times R_i \ni (x, y) \mapsto \{x, y\} \in R_i$$

is a Lipschitz map with Lipschitz inverse. Therefore, letting

$$W_i^s = \pi_i(V^s(x)) \cap R_i \quad \text{and} \quad W_i^u = \pi_i(V^u(x)) \cap R_i,$$

we obtain

$$\dim_H R_i = \dim_H \{W_i^s, W_i^u\} = \dim_H (W_i^s \times W_i^u) \tag{5.21}$$

for every $x \in R_i$, with analogous identities for the lower and upper box dimensions. By Theorem 5.1, we have

$$\dim_H W_i^s = \underline{\dim}_B W_i^s = \overline{\dim}_B W_i^s = t_s \tag{5.22}$$

and

$$\dim_H W_i^u = \underline{\dim}_B W_i^u = \overline{\dim}_B W_i^u = t_u. \tag{5.23}$$

Indeed, $V^s(x) \cap \Lambda$ is taken onto $\bigcup_{i=1}^k W_i^s$ by a map that together with its inverse are Lipschitz in each domain of continuity (compare with (5.8)). Hence, it follows from (5.21) (and the analogous identities for the lower and upper box dimensions) together with (5.22) and (5.23) that

$$\dim_H R_i = \underline{\dim}_B R_i = \overline{\dim}_B R_i = t_s + t_u. \tag{5.24}$$

Since Λ is locally diffeomorphic to the product $(-\varepsilon, \varepsilon) \times \bigcup_{i=1}^k R_i$, for any sufficiently small $\varepsilon > 0$, identity (5.20) follow readily from (5.24). □

Chapter 6
Pointwise Dimension and Applications

In this chapter, again for conformal hyperbolic flows, we establish an explicit formula for the pointwise dimension of an arbitrary invariant measure in terms of the local entropy and the Lyapunov exponents. In particular, this formula allows us to show that the Hausdorff dimension of a (nonergodic) invariant measure is equal to the essential supremum of the Hausdorff dimensions of the measures in each ergodic decomposition. We also discuss the problem of the existence of invariant measures of maximal dimension. These are measures at which the supremum of the Hausdorff dimensions over all invariant measures is attained.

6.1 A Formula for the Pointwise Dimension

In this section we consider hyperbolic flows and we establish a formula for the pointwise dimension of an arbitrary invariant measure. As a consequence, we also obtain a formula for the Hausdorff dimension of the measure.

Let $\Phi = \{\varphi_t\}_{t\in\mathbb{R}}$ be a $C^{1+\delta}$ flow in a smooth manifold M and let $\Lambda \subset M$ be a locally maximal hyperbolic set for Φ. We always assume in this chapter that the flow Φ is conformal on Λ (see Definition 5.1). Let also μ be a Φ-invariant probability measure on Λ. By Birkhoff's ergodic theorem, the limits

$$\lambda_s(x) = \lim_{t\to+\infty} \frac{1}{t} \log\|d_x\varphi_t|E^s(x)\| \quad\text{and}\quad \lambda_u(x) = \lim_{t\to+\infty} \frac{1}{t} \log\|d_x\varphi_t|E^u(x)\| \tag{6.1}$$

exist for μ-almost every $x \in \Lambda$. These are respectively the negative and positive values of the Lyapunov exponent

$$\lambda(x, v) = \limsup_{t\to+\infty} \frac{1}{t} \log \|d_x\varphi_t v\|$$

L. Barreira, *Dimension Theory of Hyperbolic Flows*,
Springer Monographs in Mathematics, DOI 10.1007/978-3-319-00548-5_6,

for $x \in \Lambda$ and $v \in T_x M$ (at μ-almost every point). On the other hand, by the Brin–Katok formula for flows, we have

$$h_\mu(x) = \lim_{\varepsilon \to 0} \lim_{t \to \infty} -\frac{1}{t} \log \mu(B(x,t,\varepsilon)) \tag{6.2}$$

for μ-almost every $x \in \Lambda$, where

$$B(x,t,\varepsilon) = \{y \in M : d(\varphi_\tau(y), \varphi_\tau(x)) < \varepsilon \text{ for } \tau \in [0,t]\}.$$

The number $h_\mu(x)$ is called the *local entropy* of μ at x. Moreover, the function $x \mapsto h_\mu(x)$ is μ-integrable and Φ-invariant μ-almost everywhere, and

$$h_\mu(\Phi) = \int_\Lambda h_\mu(x)\, d\mu(x). \tag{6.3}$$

Now we present an explicit formula for the pointwise dimension of μ in terms of the local entropy and the Lyapunov exponents.

Theorem 6.1 [21] *Let Φ be a $C^{1+\delta}$ flow with a locally maximal hyperbolic set Λ such that $\Phi|\Lambda$ is conformal and let μ be a Φ-invariant probability measure on Λ. For μ-almost every $x \in \Lambda$, we have*

$$\underline{d}_\mu(x) = \overline{d}_\mu(x) = h_\mu(x)\left(\frac{1}{\lambda_u(x)} - \frac{1}{\lambda_s(x)}\right) + 1. \tag{6.4}$$

Proof Once more, the idea of the proof is to reduce the problem to a Markov system, and then use the conformality of $\Phi|\Lambda$ to show that there exists a Moran cover of finite multiplicity. Other than this more technical aspect, all the remaining arguments are at the foundational level of ergodic theory.

Let $R_1, \ldots, R_k$ be a Markov system for Φ on Λ. For each $x \in \Lambda$ and $n \in \mathbb{N}$, we define

$$\tau_n(x) = \sum_{k=0}^{n-1} \tau(T^k(x)).$$

By Birkhoff's ergodic theorem, the limit

$$\chi(x) = \lim_{n \to \infty} \frac{\tau_n(x)}{n} \tag{6.5}$$

exists for μ-almost every $x \in \Lambda$. It follows from (3.7) that $\chi(x) > 0$ for μ-almost every $x \in \Lambda$. Now let η be the measure induced by μ on the set $Z = \bigcup_{i=1}^k R_i$. It follows from Proposition 2.3 and (4.8) that

$$h_\mu(y) = \lim_{\varepsilon \to 0} \lim_{t \to \infty} -\frac{1}{t} \log \mu(B(y,t,\varepsilon))$$

$$
\begin{aligned}
&= \lim_{\varepsilon\to 0}\lim_{n\to\infty} -\frac{1}{\tau_n(x)}\log\mu(B(y,\tau_n(x),\varepsilon)) \\
&= \frac{1}{\chi(x)}\lim_{\varepsilon\to 0}\lim_{n\to\infty} -\frac{1}{n}\log[\eta(B_Z(x,n,\varepsilon))2\varepsilon] \\
&= \frac{1}{\chi(x)}\lim_{\varepsilon\to 0}\lim_{n\to\infty} -\frac{1}{n}\log\eta(B_Z(x,n,\varepsilon))
\end{aligned} \tag{6.6}
$$

for μ-almost every $y=\varphi_s(x)\in\Lambda$, with $x\in Z$ and $s\in[0,\tau(x)]$.

Given $i_{-m},\dots,i_n\in\{1,\dots,k\}$, we define the *rectangle*

$$
R_{i_{-m}\cdots i_n}=\bigl\{x\in Z: T^j(x)\in R_{i_j}\text{ for } j=-m,\dots,n\bigr\}, \tag{6.7}
$$

where T is the transfer map in (3.5). By (6.6) and the Shannon–McMillan–Breiman theorem, for μ-almost every $x\in\Lambda$ we have

$$
h_\mu(x)=\frac{1}{\chi(x)}\lim_{n,m\to\infty} -\frac{1}{n+m}\log\eta(R_{n,m}(x)), \tag{6.8}
$$

where $R_{n,m}(x)=R_{i_{-m}\cdots i_n}$ is any rectangle such that $x\in R_{n,m}(x)$. We assume that for each $x\in\Lambda$ a particular choice of rectangles $R_{n,m}(x)$ is made from the beginning, for all $n,m\in\mathbb{N}$. Let $X\subset\Lambda$ be a full μ-measure Φ-invariant set such that for each $x\in\Lambda$:

1. the numbers $\lambda_s(x)$ and $\lambda_u(x)$ in (6.1) and $\chi(x)$ in (6.5) are well defined;
2. the number $h_\mu(x)$ in (6.2) is well defined and identity (6.8) holds.

We proceed with the proof of the theorem. Take $\varepsilon>0$. For each $x\in X\setminus Z$, there exists a $p(x)\in\mathbb{N}$ such that if $t\ge p(x)$, then

$$
\lambda_s(x)-\varepsilon<\frac{1}{t}\log\|d_x\varphi_t|E^s(x)\|<\lambda_s(x)+\varepsilon, \tag{6.9}
$$

$$
\lambda_u(x)-\varepsilon<\frac{1}{t}\log\|d_x\varphi_t|E^u(x)\|<\lambda_u(x)+\varepsilon, \tag{6.10}
$$

and if $n,m\ge p(x)$, then

$$
\chi(x)-\varepsilon<\frac{\tau_n(x)}{n}<\chi(x)+\varepsilon, \tag{6.11}
$$

$$
-h_\mu(x)\chi(x)-\varepsilon<\frac{1}{n+m}\log\eta(R_{n,m}(x))<-h_\mu(x)\chi(x)+\varepsilon. \tag{6.12}
$$

Given $\ell\in\mathbb{N}$, we consider the set

$$
Q_\ell=\bigl\{x\in X: p(x)\le\ell\bigr\}.
$$

Clearly, $\bigcup_{\ell\in\mathbb{N}} Q_\ell = X$. For each $x \in X$, there exists an $r(x) > 0$ such that for any $r \in (0, r(x))$ one can choose integers $m = m(x,r)$ and $n = n(x,r)$ with $\tau_m(x), \tau_n(x) \geq p(x)$ for which

$$\|d_x\varphi_{\tau_m(x)}|E^s(x)\| \geq r, \quad \|d_x\varphi_{\tau_{m+1}(x)}|E^s(x)\| < r \tag{6.13}$$

and

$$\|d_x\varphi_{\tau_n(x)}|E^u(x)\|^{-1} \geq r, \quad \|d_x\varphi_{\tau_{n+1}(x)}|E^u(x)\|^{-1} < r. \tag{6.14}$$

Combining (6.9) with (6.13), and (6.10) with (6.14), we obtain

$$\tau_m(x)(\lambda_s(x) - \varepsilon) < \log r + a, \quad \log r < \tau_m(x)(\lambda_s(x) + \varepsilon) \tag{6.15}$$

and

$$-\log r - a < \tau_n(x)(\lambda_u(x) + \varepsilon), \quad \tau_n(x)(\lambda_u(x) - \varepsilon) < -\log r, \tag{6.16}$$

where

$$a = \max\left\{-\inf_{x\in Z} \log\|d_x\varphi_{\tau(x)}|E^s(x)\|, \sup_{x\in Z} \log\|d_x\varphi_{\tau(x)}|E^u(x)\|\right\}.$$

We write $R(x,r) = R_{n(x,r),m(x,r)}(x)$.

We first establish an upper bound for the pointwise dimension. By Proposition 2.3 and the conformality of Φ on Λ, there exists a $c > 0$ (independent of x and r) such that

$$B(x, cr) \supset R(x,r) \times I_r(x),$$

where $I_r(x)$ is some interval of length $2r$. By (4.8), (6.11) and (6.12), for each $x \in X \setminus Z$ and any sufficiently small r, we obtain

$$\begin{aligned}
\mu(B(x,cr)) &\geq \eta(R(x,r))2r \\
&\geq \exp[(-h_\mu(x)\chi(x) - \varepsilon)(n+m)]2r \\
&\geq \exp[-h_\mu(x)(\tau_n(x) + \tau_m(x)) - (h_\mu(x)+1)\varepsilon(n+m)]2r \\
&\geq \exp[(-h_\mu(x) - (h_\mu(x)+1)\varepsilon/\sigma)(\tau_n(x) + \tau_m(x))]2r,
\end{aligned}$$

where $\sigma = \inf_Z \tau > 0$ (see (3.7)). Using (6.15) and (6.16), we conclude that

$$\mu(B(x,cr)) \geq \exp\left[\left(h_\mu(x) + (h_\mu(x)+1)\frac{\varepsilon}{\sigma}\right)\left(\frac{\log r}{\lambda_u(x) - \varepsilon} - \frac{\log r}{\lambda_s(x) + \varepsilon}\right)\right] 2r.$$

Taking logarithms and letting $r \to 0$, we finally obtain

$$\overline{d}_\mu(x) \leq \left(h_\mu(x) + (h_\mu(x)+1)\frac{\varepsilon}{\sigma}\right)\left(\frac{1}{\lambda_u(x) - \varepsilon} - \frac{1}{\lambda_s(x) + \varepsilon}\right) + 1.$$

The arbitrariness of ε implies that

$$\overline{d}_\mu(x) \le h_\mu(x)\left(\frac{1}{\lambda_u(x)} - \frac{1}{\lambda_s(x)}\right) + 1$$

for every $x \in X$ and hence for μ-almost every $x \in \Lambda$.

Now we establish a lower bound for the pointwise dimension. Take $\varepsilon > 0$. Given $x \in X$, we define

$$\begin{aligned}\Gamma(x) = \{y \in X : |\lambda_s(y) - \lambda_s(x)| < \varepsilon, |\lambda_u(y) - \lambda_u(x)| < \varepsilon,\\ |\chi(y) - \chi(x)| < \varepsilon, \text{ and } |h_\mu(y) - h_\mu(x)| < \varepsilon\}. \quad (6.17)\end{aligned}$$

We note that the sets $\Gamma(x)$ are Φ-invariant. Moreover, they cover X and one can choose points $y_i \in X$ for $i = 1, 2, \ldots$ such that $\Gamma_i = \Gamma(y_i)$ has measure $\mu(\Gamma_i) > 0$ for each i, and $\bigcup_{i\in\mathbb{N}} \Gamma_i$ has full μ-measure.

Take $i, \ell \in \mathbb{N}$. We proceed in a similar manner to that in [81, Sect. 22] to construct a cover of $\Gamma_i \cap Q_\ell \cap Z$ by sets $R(x,r)$ (we recall that $Z = \bigcup_{i=1}^k R_i$). For each $x \in \Gamma_i \cap Q_\ell$ and $r > 0$, we denote by $R'(x,r)$ the largest rectangle containing x (among those in (6.7)) with the property that

$$R'(x,r) = R(y,r) \quad \text{for some} \quad y \in R'(x,r) \cap \Gamma_i \cap Q_\ell \cap Z,$$

and

$$R(z,r) \subset R'(x,r) \quad \text{for every} \quad z \in R'(x,r) \cap \Gamma_i \cap Q_\ell \cap Z.$$

Two sets $R'(x,r)$ and $R'(y,r)$ either coincide or intersect at most along their boundaries. It follows from the Borel density lemma (see [42, Theorem 2.9.11]) that for μ-almost every $x \in \Gamma_i \cap Q_\ell$ there exists an $r(x) > 0$ such that

$$\mu(B(x,r)) \le 2\mu(B(x,r) \cap \Gamma_i \cap Q_\ell)$$

for every $r \in (0, r(x))$.

Again by the conformality of Φ on Λ and the uniform transversality of the stable and unstable manifolds, there exist a $K > 0$ (independent of x and r) and points $x_1, \ldots, x_k \in \Gamma_i \cap Q_\ell$ with $k \le K$ such that

$$B(x,r) \cap \Gamma_i \cap Q_\ell \subset \bigcup_{j=1}^k \big(R'(x_j,r) \times I_r(x_j)\big).$$

Therefore,

$$
\begin{aligned}
\mu(B(x,r)) &\leq 2\mu(B(x,r)\cap \Gamma_i \cap Q_\ell)\\
&\leq 4r\sum_{j=1}^{k}\mu(R'(x_j,r))\\
&\leq 4r\sum_{j=1}^{k}\exp[(-h_\mu(x_j)\chi(x_j)+\varepsilon)(n(x_j,r)+m(x_j,r))]\\
&\leq 4r\sum_{j=1}^{k}\exp\Big[\Big(-h_\mu(x_j)+(h_\mu(x_j)+1)\frac{\varepsilon}{\sigma}\Big)(\tau_n(x_j)+\tau_m(x_j))\Big],
\end{aligned}
$$

using (6.12). By (6.15), (6.16) and the definition of Γ_i, we conclude that

$$
\begin{aligned}
\mu(B(x,r)) &\leq 4r\sum_{j=1}^{k}\exp\left[a(y_i)\left(\frac{\log r+a}{\lambda_u(x_j)+\varepsilon}-\frac{\log r+a}{\lambda_s(x_j)-\varepsilon}\right)\right]\\
&\leq 4rK\exp\left[b(x)\left(\frac{\log r+a}{\lambda_u(x)+2\varepsilon}-\frac{\log r+a}{\lambda_s(x)-2\varepsilon}\right)\right],
\end{aligned}
$$

where

$$a(y_i)=h_\mu(y_i)-\varepsilon-(h_\mu(y_i)+\varepsilon+1)\varepsilon/\sigma$$

and

$$b(x)=h_\mu(x)-2\varepsilon-(h_\mu(x)+2\varepsilon+1)\varepsilon/\sigma.$$

Taking logarithms and letting $r\to 0$ we find that

$$\underline{d}_\mu(x)\geq\Big(h_\mu(x)-2\varepsilon-(h_\mu(x)+2\varepsilon+1)\frac{\varepsilon}{\sigma}\Big)\left(\frac{1}{\lambda_u(x)+2\varepsilon}-\frac{1}{\lambda_s(x)-2\varepsilon}\right)+1$$

for μ-almost every $x\in\Gamma_i\cap Q_\ell$. Letting $\ell\to\infty$, we conclude that this inequality holds for μ-almost every $x\in\Gamma_i$, and the arbitrariness of ε implies that

$$\underline{d}_\mu(x)\geq h_\mu(x)\left(\frac{1}{\lambda_u(x)}-\frac{1}{\lambda_s(x)}\right)+1, \tag{6.18}$$

also for μ-almost every$x\in\Gamma$. Since $\bigcup_{\ell\in\mathbb{N}}\Gamma_i$ has full μ-measure, inequality (6.18) holds for μ-almost every $x\in\Lambda$. This completes the proof of the theorem. □

In [82], Pesin and Sadovskaya established the identities in (6.4) for equilibrium measures for a Hölder continuous function. We note that these measures are ergodic and have a local product structure, while Theorem 6.1 considers arbitrary invariant measures.

The following result is a simple consequence of Theorem 6.1.

Theorem 6.2 *If Φ is a $C^{1+\delta}$ flow with a locally maximal hyperbolic set Λ such that $\Phi|\Lambda$ is conformal and μ is a Φ-invariant probability measure on Λ, then*

$$\dim_H \mu = \operatorname{ess\,sup}\left\{h_\mu(x)\left(\frac{1}{\lambda_u(x)} - \frac{1}{\lambda_s(x)}\right) + 1 : x \in \Lambda\right\}. \tag{6.19}$$

If, in addition, μ is ergodic, then

$$\dim_H \mu = h_\mu(\Phi)\left(\frac{1}{\int_\Lambda \zeta_u \, d\mu} - \frac{1}{\int_\Lambda \zeta_s \, d\mu}\right) + 1. \tag{6.20}$$

Proof Combining Theorem 6.1 with Proposition 4.7 yields identity (6.19). When μ is ergodic, since the functions h_μ, λ_s and λ_u are Φ-invariant they are constant μ-almost everywhere. It follows from (6.3) that $h_\mu(x) = h_\mu(\Phi)$ for μ-almost every $x \in \Lambda$. Moreover, proceeding as in (5.7), we have

$$\int_\Lambda \zeta_u \, d\mu = \int_\Lambda \lambda_u(x) \, d\mu(x) \quad \text{and} \quad \int_\Lambda \zeta_s \, d\mu = \int_\Lambda \lambda_s(x) \, d\mu(x).$$

Hence,

$$\lambda_u(x) = \int_\Lambda \zeta_u \, d\mu \quad \text{and} \quad \lambda_s(x) = \int_\Lambda \zeta_s \, d\mu$$

for μ-almost every $x \in \Lambda$. Therefore,

$$h_\mu(x)\left(\frac{1}{\lambda_u(x)} - \frac{1}{\lambda_s(x)}\right) + 1 = h_\mu(\Phi)\left(\frac{1}{\int_\Lambda \zeta_u \, d\mu} - \frac{1}{\int_\Lambda \zeta_s \, d\mu}\right) + 1,$$

also for μ-almost every $x \in \Lambda$. Identity (6.20) now follows readily from Theorem 6.1 together with Proposition 4.6. □

6.2 Hausdorff Dimension and Ergodic Decompositions

In this section we describe the behavior of the Hausdorff dimension of an invariant measure with respect to an ergodic decomposition.

Theorem 6.3 ([21]) *Let Φ be a $C^{1+\delta}$ flow with a locally maximal hyperbolic set Λ such that $\Phi|\Lambda$ is conformal and let μ be a Φ-invariant probability measure on Λ. For each ergodic decomposition τ of μ, we have*

$$\dim_H \mu = \operatorname{ess\,sup}\{\dim_H \nu : \nu \in \mathcal{M}_E\},$$

with the essential supremum taken with respect to τ.

Proof By Proposition 4.8, we have

$$\dim_H \mu \geq \operatorname{ess\,sup}\{\dim_H \nu : \nu \in \mathcal{M}_E\}.$$

Now we establish the reverse inequality. Take $\varepsilon > 0$. For each x in the full μ-measure set X considered in the proof of Theorem 6.1, let $\Gamma(x)$ be the set in (6.17). We also take points $y_i \in X$ for $i = 1, 2, \ldots$ such that the Φ-invariant sets $\Gamma_i = \Gamma(y_i)$ have measure $\mu(\Gamma_i) > 0$ for each i, and $\bigcup_{i\in\mathbb{N}} \Gamma_i$ has full μ-measure. For each i we consider the normalized restriction μ_i of μ to Γ_i. It follows from (6.3) and (6.17) that

$$h_{\mu_i}(\Phi|\Gamma_i) = \frac{1}{\mu(\Gamma_i)} \int_{\Gamma_i} h_\mu(x)\, d\mu(x) \geq h_\mu(y_i) - \varepsilon. \tag{6.21}$$

We note that a measure $\nu \in \mathcal{M}$ is ergodic (with respect to Φ) if and only if the induced measure η_ν on Z is ergodic (with respect to T). Moreover, the ergodic decomposition τ induces a measure τ_Z on the set $\mathcal{M}_Z$ of all T-invariant probability measures on Z. We consider a new measure $\tilde{\tau}_Z$ on $\mathcal{M}_Z$ with Radon–Nikodym derivative

$$\frac{d\tilde{\tau}_Z}{d\tau_Z}(\eta) = \frac{\int_Z \tau\, d\eta_\mu}{\int_Z \tau\, d\eta}. \tag{6.22}$$

Now let $G\colon Z \to \mathbb{R}$ be a continuous function. We define another function $g\colon \Lambda \to \mathbb{R}$ by $g(\varphi_s(x)) = G(x)/\tau(x)$ for $x \in Z$ and $s \in [0, \tau(x))$. Clearly,

$$G(x) = \int_0^{\tau(x)} g(\varphi_s(x))\, ds,$$

and we have

$$\begin{aligned}
\int_{\mathcal{M}_Z} \int_Z G\, d\eta d\tilde{\tau}_Z(\eta) &= \int_Z \tau\, d\eta_\mu \int_{\mathcal{M}_Z} \frac{\int_Z G\, d\eta}{\int_Z \tau\, d\eta}\, d\tau_Z(\eta) \\
&= \int_Z \tau\, d\eta_\mu \int_{\mathcal{M}_Z} \int_\Lambda g\, d\nu\, d\tau(\nu) \\
&= \int_Z \tau\, d\eta_\mu \int_\Lambda g\, d\mu = \int_Z G\, d\eta_\mu.
\end{aligned}$$

This shows that $\tilde{\tau}_Z$ is an ergodic decomposition of η_μ. Since the Radon–Nikodym derivative in (6.22) is bounded and bounded away from zero, a subset of $\mathcal{M}_Z$ has positive τ_Z-measure if and only if it has positive $\tilde{\tau}_Z$-measure.

Now let

$$\mathcal{M}_i = \{\nu \in \mathcal{M} : \nu(\Lambda \setminus \Gamma_i) = 0\}.$$

Since Γ_i is Φ-invariant, there exists a one-to-one correspondence between the ergodic Φ-invariant probability measures on Γ_i and the measures in $\mathcal{M}_i \cap \mathcal{M}_E$. Therefore, $\tau(\mathcal{M}_i \cap \mathcal{M}_E) > 0$ and the normalized restriction τ_i of τ to $\mathcal{M}_i$ is an ergodic decomposition of $\mu|\Gamma_i$ with respect to Φ. We have

$$h_{\eta_i}(T|\Gamma_i \cap Z) = \int_{\mathcal{M}_i} h_\eta(T)\, d\tau_i(\nu),$$

where η and η_i are the measures induced respectively by ν and μ_i on Z. Hence, there exists a set $A_i \subset \mathcal{M}_i \cap \mathcal{M}_E$ of positive τ_i-measure, and thus also of positive τ-measure, such that

$$h_\eta(T) > h_{\eta_i}(T|\Gamma_i \cap Z) - \varepsilon$$

for each $\nu \in A_i$. Since

$$\int_{\mathcal{M}_i} \int_{\Gamma_i \cap Z} \tau\, d\eta\, d\tilde{\tau}_i(\eta) = \int_{\Gamma_i \cap Z} \tau\, d\mu_i,$$

one can also assume that

$$\int_{\Gamma_i \cap Z} \tau\, d\eta \le \int_{\Gamma_i \cap Z} \tau\, d\eta_i + \varepsilon \tag{6.23}$$

for every $\eta_\nu \in \mathcal{M}_Z$ with $\nu \in A_i$. Therefore, for each $\nu \in A_i$ and $x \in \Gamma_i$, we have

$$\begin{aligned} \frac{h_\eta(T) + \varepsilon}{\int_{\Gamma_i \cap Z} \tau\, d\eta - \varepsilon} &\ge \frac{h_{\mu_i}(T|\Gamma_i \cap Z)}{\int_{\Gamma_i \cap Z} \tau\, d\eta_i} = h_{\mu_i}(\Phi|\Gamma_i) \\ &\ge h_\mu(y_i) - 2\varepsilon > h_\mu(x) - 3\varepsilon, \end{aligned} \tag{6.24}$$

using (6.21). On the other hand, for $\nu \in A_i$ and $x \in \Gamma_i$, we have

$$\int_\Lambda \zeta_s\, d\nu \ge \lambda_s(y_i) - \varepsilon \ge \lambda_s(x) - 2\varepsilon$$

and

$$\int_\Lambda \zeta_u\, d\nu \le \lambda_u(y_i) + \varepsilon \le \lambda_u(x) + 2\varepsilon.$$

For each $x \in X$, combining (6.23) and (6.24) with (6.20), we conclude that

$$\begin{aligned} h_\mu(x)\left(\frac{1}{\lambda_u(x)} - \frac{1}{\lambda_s(x)}\right) + 1 &\le \left(\frac{h_\eta(T) + \varepsilon}{\int_{\Gamma_i \cap Z} \tau\, d\eta - \varepsilon} + 3\varepsilon\right) \\ &\quad \times \left(\frac{1}{\int_\Lambda \zeta_u\, d\nu - 2\varepsilon} - \frac{1}{\int_\Lambda \zeta_s\, d\nu + 2\varepsilon}\right) + 1 \\ &\le \dim_H \nu + C(\varepsilon) \end{aligned}$$

for every $\nu \in A_i$, where C is a function (independent of i and ν) such that $C(\varepsilon) \to 0$ when $\varepsilon \to 0$. Since A_i has positive τ-measure, it follows from (6.19) that

$$\dim_H \mu \le \operatorname{ess\,sup}\{\dim_H \nu : \nu \in \mathcal{M}_E\} + C(\varepsilon),$$

and the arbitrariness of ε yields the desired result. □

6.3 Measures of Maximal Dimension

In this section we establish the existence of measures of maximal dimension on a locally maximal hyperbolic set for a conformal flow.

We first introduce the notion of a measure of maximal dimension. Let Φ be a C^1 flow with a locally maximal hyperbolic set Λ.

Definition 6.1 A measure $\mu \in \mathcal{M}$ such that

$$\dim_H \mu = \sup\{\dim_H \nu : \nu \in \mathcal{M}\}$$

is called a *measure of maximal dimension*.

Now we assume that $\Phi|\Lambda$ is conformal. For each $\nu \in \mathcal{M}$, let

$$\lambda_s(\nu) = \int_\Lambda \lambda_s(x)\, d\nu(x) \quad \text{and} \quad \lambda_u(\nu) = \int_\Lambda \lambda_u(x)\, d\nu(x),$$

with $\lambda_s(x)$ and $\lambda_u(x)$ as in (6.1). Since

$$\lambda_s(x) = \lim_{t\to+\infty} \frac{1}{t}\int_0^t \zeta_s(\varphi_\tau(x))\, d\tau \quad \text{and} \quad \lambda_u(x) = \lim_{t\to+\infty} \frac{1}{t}\int_0^t \zeta_u(\varphi_\tau(x))\, d\tau,$$

it follows from Birkhoff's ergodic theorem that

$$\lambda_s(\nu) = \int_\Lambda \zeta_s\, d\nu \quad \text{and} \quad \lambda_u(\nu) = \int_\Lambda \zeta_u\, d\nu. \tag{6.25}$$

The following result establishes the existence of measures of maximal dimension on a locally maximal hyperbolic set for a conformal flow.

Theorem 6.4 *If Φ is a $C^{1+\delta}$ flow with a locally maximal hyperbolic set Λ such that $\Phi|\Lambda$ is conformal and topologically mixing, then there exists an ergodic measure of maximal dimension.*

Proof The proof closely follows arguments of Barreira and Wolf in [19] for discrete time. We define a function $Q\colon \mathbb{R}^2 \to \mathbb{R}$ by

$$Q(p,q) = P_{\Phi|\Lambda}(-p\zeta_u + q\zeta_s).$$

By Proposition 4.4, since ζ_s and ζ_u are Hölder continuous, the function Q is analytic. Moreover, for each $(p,q) \in \mathbb{R}^2$ the Hölder continuous function $-p\zeta_u + q\zeta_s$ has a unique equilibrium measure, which we denote by $\nu_{p,q}$. We also write

$$\lambda_u(p,q) = \lambda_u(\nu_{p,q}), \quad \lambda_s(p,q) = \lambda_s(\nu_{p,q}), \quad h(p,q) = h_{\nu_{p,q}}(\Phi),$$

and we think of λ_u, λ_s and h as functions on $\mathbb{R}^2$. By Proposition 4.2, we have

$$Q(p,q) = h(p,q) - p\lambda_u(p,q) + q\lambda_s(p,q). \tag{6.26}$$

Since the maps $\nu \mapsto \lambda_s(\nu)$ and $\nu \mapsto \lambda_u(\nu)$ given by (6.25) are continuous on the compact set $\mathcal{M}$ (when equipped with the weak* topology), one can define

$$\lambda_s^{\min} = \min \lambda_s(\mathcal{M}), \quad \lambda_s^{\max} = \max \lambda_s(\mathcal{M})$$

and

$$\lambda_u^{\min} = \min \lambda_u(\mathcal{M}), \quad \lambda_u^{\max} = \max \lambda_u(\mathcal{M}).$$

We also consider the intervals

$$I_s = (\lambda_s^{\min}, \lambda_s^{\max}) \quad \text{and} \quad I_u = (\lambda_u^{\min}, \lambda_u^{\max}).$$

We note that $I_s \neq \varnothing$ (respectively $I_u \neq \varnothing$) if and only if ζ_s (respectively ζ_u) is not cohomologous to a constant.

Now we consider the functions

$$d_u(p,q) = h(p,q)/\lambda_u(p,q) \quad \text{and} \quad d_s(p,q) = -h(p,q)/\lambda_s(p,q). \tag{6.27}$$

It follows from (4.6) that

$$\frac{\partial Q}{\partial p} = -\lambda_u \quad \text{and} \quad \frac{\partial Q}{\partial q} = \lambda_s. \tag{6.28}$$

Since Q is analytic, the functions λ_u and λ_s are also analytic. We conclude from (6.26) that h is analytic, and it follows from (6.27) that the functions d_u and d_s are also analytic.

Lemma 6.1 *The following properties hold*:

1. *if ζ_u is not cohomologous to a constant, then for each $q \in \mathbb{R}$*:
   a. *$\lambda_u(\cdot, q)$ is strictly decreasing and $\{\lambda_u(p,q) : p \in \mathbb{R}\} = I_u$;*
   b. *$h(\cdot, 0)$ is strictly decreasing in $[0, \infty)$;*
   c. *$d_u(\cdot, 0)$ is strictly increasing in $(-\infty, t_u]$ and strictly decreasing in $[t_u, \infty)$.*
2. *if ζ_s is not cohomologous to a constant, then for each $p \in \mathbb{R}$*:
   a. *$\lambda_s(p, \cdot)$ is strictly decreasing and $\{\lambda_s(p,q) : q \in \mathbb{R}\} = I_s$;*
   b. *$h(0, \cdot)$ is strictly decreasing in $[0, \infty)$;*
   c. *$d_s(0, \cdot)$ is strictly increasing in $(-\infty, t_s]$ and strictly decreasing in $[t_s, \infty)$.*

Proof of the lemma Let us assume that ζ_u is not cohomologous to a constant and take $q \in \mathbb{R}$. By (4.7) and (6.28), we have

$$\frac{\partial \lambda_u}{\partial p} = -\frac{\partial^2 Q}{\partial p^2} < 0, \tag{6.29}$$

and thus $\lambda_u(\cdot, q)$ is strictly decreasing. Since the function $\lambda_u(\cdot, q)$ is continuous, the set $\{\lambda_u(p,q) : p \in \mathbb{R}\}$ is an open interval. We claim that

$$\lim_{p \to \infty} \lambda_u(p,q) = \lambda_u^{\min} \quad \text{and} \quad \lim_{p \to -\infty} \lambda_u(p,q) = \lambda_u^{\max}. \tag{6.30}$$

If the first identity did not hold, then there would exist $\nu \in \mathcal{M}$ and $\delta > 0$ such that $\lambda_u(\nu) + \delta < \lambda_u(p,q)$ for $p \in \mathbb{R}$. Now take $p > 0$ satisfying

$$p\delta > h(\Phi|\Lambda) - q\lambda_s(\nu) + q\lambda_s(p,q)$$

(such a p always exists, because the function $\lambda_s(\cdot, q)$ is bounded). We obtain

$$\begin{aligned} Q(p,q) &= h(p,q) - p\lambda_u(p,q) + q\lambda_s(p,q) \\ &< h(\Phi|\Lambda) - p(\lambda_u(\nu) + \delta) + q\lambda_s(p,q) \\ &< h_\nu(\Phi) - p\lambda_u(\nu) + q\lambda_s(\nu), \end{aligned}$$

which contradicts Proposition 4.2. This establishes the first identity in (6.30). A similar argument establishes the second identity.

It follows from (6.26) that

$$h(p,0) = Q(p,0) + p\lambda_u(p,0).$$

Using (6.28) and (6.29), one can easily verify that

$$\frac{\partial h}{\partial p}(p,0) = p\frac{\partial \lambda_u}{\partial p}(p,0). \tag{6.31}$$

This establishes property 1b.

Finally, using (6.26), (6.29) and (6.31), we obtain

$$\begin{aligned} \frac{\partial d_u}{\partial p}(p,0) &= \frac{p\partial\lambda_u/\partial p(p,0)\lambda_u(p,0) - h(p,0)\partial\lambda_u/\partial p(p,0)}{\lambda_u(p,0)^2} \\ &= -\frac{\partial \lambda_u}{\partial p}(p,0)\frac{Q(p,0)}{\lambda_u(p,0)^2} \\ &= \frac{\partial^2 Q}{\partial p^2}(p,0)\frac{Q(p,0)}{\lambda_u(p,0)^2}. \end{aligned} \tag{6.32}$$

On the other hand, it follows from Proposition 4.2 that the function $Q(\cdot, q)$ is strictly decreasing. This implies that

$$Q(p,0) > Q(t_u,0) = 0 \quad \text{for} \quad p < t_u,$$

and

$$Q(p,0) < Q(t_u,0) = 0 \quad \text{for} \quad p > t_u.$$

Property 1c follows now immediately from (6.29) and (6.32).

The proofs of the remaining statements are analogous. □

Using Lemma 6.1 one can introduce two curves that are crucial to our approach.

Lemma 6.2 *The following properties hold:*

1. *for each $a \in I_u$ there exists a unique function $\gamma_u \colon \mathbb{R} \to \mathbb{R}$ satisfying*

$$\lambda_u(\gamma_u(q), q) = a \quad \textit{for} \quad q \in \mathbb{R},$$

and γ_u is analytic;
2. *for each $b \in I_s$ there exists a unique function $\gamma_s \colon \mathbb{R} \to \mathbb{R}$ satisfying*

$$\lambda_s(p, \gamma_s(p)) = b \quad \textit{for} \quad p \in \mathbb{R},$$

and γ_s is analytic.

Proof of the lemma We only prove the second statement. The proof of the first statement is analogous. Let $b \in I_s$. In particular, $I_s \neq \varnothing$ and ζ_s is not cohomologous to a constant. By statement 2a in Lemma 6.1 and (6.28), for each $p \in \mathbb{R}$ there exists a unique number $\gamma_s(p) \in \mathbb{R}$ such that

$$\frac{\partial Q}{\partial q}(p, \gamma_s(p)) = \lambda_s(p, \gamma_s(p)) = b.$$

Since ζ_s is not cohomologous to a constant, we have $\partial^2 Q/\partial q^2(p,q) > 0$ for every $(p,q) \in \mathbb{R}^2$ and it follows from the Implicit function theorem that the map $p \mapsto \gamma_s(p)$ is analytic. □

We proceed with the proof of the theorem. Let $(\nu_n)_{n\in\mathbb{N}}$ be a sequence of measures in $\mathcal{M}_E$ (that is, a sequence of ergodic measures in $\mathcal{M}$) such that

$$\lim_{n\to\infty} \dim_H \nu_n = \sup\{\dim_H \nu : \nu \in \mathcal{M}_E\}. \tag{6.33}$$

Since $\mathcal{M}$ is compact, one can also assume that $(\nu_n)_{n\in\mathbb{N}}$ converges to some measure $m \in \mathcal{M}$. Since the map $\mathcal{M} \ni \nu \mapsto h_\nu(\Phi)$ is upper semicontinuous, it follows from (6.19) and the continuity of the maps $\nu \mapsto \lambda_u(\nu)$ and $\nu \mapsto \lambda_s(\nu)$ that

$$\lim_{n\to\infty} \dim_H \nu_n \le d(m), \tag{6.34}$$

where

$$d(m) = h_m(\Phi)\left(\frac{1}{\int_\Lambda \zeta_u\,dm} - \frac{1}{\int_\Lambda \zeta_s\,dm}\right) + 1.$$

By (6.33) and (6.34), we obtain

$$\sup\{\dim_H \nu : \nu \in \mathcal{M}_E\} \le d(m). \tag{6.35}$$

Hence, in order to establish the existence of a measure $\mu \in \mathcal{M}_E$ satisfying

$$\dim_H \mu = \sup\{\dim_H \nu : \nu \in \mathcal{M}_E\} \tag{6.36}$$

it is sufficient to show that there exists a $\mu \in \mathcal{M}_E$ such that

$$\dim_H \mu = d(m). \tag{6.37}$$

Clearly, any measure $\mu \in \mathcal{M}_E$ satisfying (6.37) also satisfies (6.36). When m is ergodic, it follows from (6.19) that $\dim_H m = d(m)$, and hence identity (6.36) holds for $\mu = m$. However, the measure m may not be ergodic.

Let $a = \lambda_u(m)$ and $b = \lambda_s(m)$. By Lemma 6.2, when $a \in I_u$ (respectively $b \in I_s$) one can consider the curve γ_u (respectively γ_s) associated to the number a (respectively b). Now we prove some auxiliary statements.

Lemma 6.3 *If $\lambda_s(m) \in I_s$, then there exists a $p \in [0, h_m(\Phi)/\lambda_u(m)]$ such that*

$$\lambda_u(p, \gamma_s(p)) = \lambda_u(m).$$

Proof of the lemma The assumption $\lambda_s(m) \in I_s$ guarantees that the function γ_s is well defined. Since $\nu_{p,\gamma_s(p)}$ is the equilibrium measure of $-p\zeta_u + \gamma_s(p)\zeta_s$, we have

$$h(p, \gamma_s(p)) - p\lambda_u(p, \gamma_s(p)) + \gamma_s(p)\lambda_s(p, \gamma_s(p)) \geq h_m(\Phi) - p\lambda_u(m) + \gamma_s(p)\lambda_s(m) \tag{6.38}$$

for $p \in \mathbb{R}$. One can easily verify that

$$\frac{h(p, \gamma_s(p))}{\lambda_u(p, \gamma_s(p))} - \frac{h_m(\Phi)}{\lambda_u(m)} \geq \left(1 - \frac{\lambda_u(m)}{\lambda_u(p, \gamma_s(p))}\right)\left(p - \frac{h_m(\Phi)}{\lambda_u(m)}\right). \tag{6.39}$$

Let $\kappa = h_m(\Phi)/\lambda_u(m)$. It follows from (6.39) with $p = \kappa$ that

$$h(\kappa, \gamma_s(\kappa))/\lambda_u(\kappa, \gamma_s(\kappa)) \geq h_m(\Phi)/\lambda_u(m). \tag{6.40}$$

Now let us assume that

$$\lambda_u(\kappa, \gamma_s(\kappa)) > \lambda_u(m).$$

By (6.40), we obtain $h(\kappa, \gamma_s(\kappa)) > h_m(\Phi)$. It follows from (6.19) and (6.40) that $\dim_H \nu_{\kappa,\gamma_s(\kappa)} > d(m)$. This contradicts (6.35), and thus, we must have

$$\lambda_u(\kappa, \gamma_s(\kappa)) \leq \lambda_u(m). \tag{6.41}$$

On the other hand, it follows from (6.19) and (6.35) that

$$\frac{h(0, \gamma_s(0))}{\lambda_u(0, \gamma_s(0))} - \frac{h(0, \gamma_s(0))}{\lambda_s(m)} \leq \frac{h_m(\Phi)}{\lambda_u(m)} - \frac{h_m(\Phi)}{\lambda_s(m)}. \tag{6.42}$$

Taking $p = 0$ in (6.38), we obtain $h(0, \gamma_s(0)) \geq h_m(\Phi)$, and it follows from (6.42) that

$$\lambda_u(0, \gamma_s(0)) \geq \lambda_u(m). \tag{6.43}$$

By the continuity of the function $p \mapsto \lambda_u(p, \gamma_u(p))$ together with (6.41) and (6.43), there exists a $p \in [0, \kappa]$ such that $\lambda_u(p, \gamma_s(p)) = \lambda_u(m)$. This completes the proof of the lemma. □

Lemma 6.4 *Assume that neither ζ_u nor ζ_s are cohomologous to a constant. Then $\lambda_u(m) \in I_u$ if and only if $\lambda_s(m) \in I_s$.*

Proof of the lemma Let us assume that $\lambda_s(m) \in I_s$. By Lemma 6.3, there exists a p such that $\lambda_u(p, \gamma_s(p)) = \lambda_u(m)$. By Lemma 6.1, we have $\lambda_u(p, \gamma_s(p)) \in I_u$ and hence $\lambda_u(m) \in I_u$. A similar argument together with the corresponding version of Lemma 6.3 show that $\lambda_s(m) \in I_s$ whenever $\lambda_u(m) \in I_u$. □

By Lemma 6.4, it is sufficient to consider four cases:

1. $\lambda_s(m) \in I_s$ and $\lambda_u(m) \in I_u$;
2. $\lambda_s(m) \in I_s$ and ζ_u is cohomologous to a constant;
3. $\lambda_u(m) \in I_u$ and ζ_s is cohomologous to a constant;
4. $\lambda_s(m) \notin I_s$ and $\lambda_u(m) \notin I_u$.

We still need another auxiliary statement.

Lemma 6.5 *If $p, q \in \mathbb{R}$ are such that*

$$\lambda_u(p, q) = \lambda_u(m) \quad \textit{and} \quad \lambda_s(p, q) = \lambda_s(m),$$

then $m = \nu_{p,q}$.

Proof of the lemma We have

$$\begin{aligned} h(p, q) + \int_\Lambda (-p\zeta_u + q\zeta_s)\, d\nu_{p,q} &= h(p, q) - p\lambda_u(m) + q\lambda_s(m) \\ &\geq h_m(\Phi) + \int_\Lambda (-p\zeta_u + q\zeta_s)\, dm. \end{aligned}$$

Hence, $h(p, q) \geq h_m(\Phi)$, with equality if and only if $\nu_{p,q} = m$. On the other hand, combining (6.19) with (6.35) we obtain $h(p, q) \leq h_m(\Phi)$. Therefore $h(p, q) = h_m(\Phi)$ and $m = \nu_{p,q}$. □

Now we consider each of the above four cases.

Lemma 6.6 *If $\lambda_u(m) \in I_u$ and $\lambda_s(m) \in I_s$, then there exist $p, q \in \mathbb{R}$ such that $(p, \gamma_s(p)) = (\gamma_u(q), q)$ and $m = \nu_{p,q}$.*

Proof of the lemma The hypotheses of the lemma guarantee that the curves γ_u and γ_s are well defined. Since $\lambda_s(p, \gamma_s(p)) = \lambda_s(m)$, it follows from Lemma 6.3 and the uniqueness of γ_u that $(p, \gamma_s(p)) = (\gamma_u(q), q)$ for some $p, q \in \mathbb{R}$. In particular,

$$\lambda_u(p, q) = \lambda_u(m) \quad \text{and} \quad \lambda_s(p, q) = \lambda_s(m).$$

Hence, it follows from Lemma 6.5 that $m = \nu_{p,q}$. □

Lemma 6.7 *If $\lambda_s(m) \in I_s$ and ζ_u is cohomologous to a constant, then there exist $p, q \in \mathbb{R}$ such that $m = \nu_{p,q}$.*

Proof of the lemma Since $\lambda_s(m) \in I_s$, the curve γ_s is well defined, and $\lambda_s(p, \gamma_s(p)) = \lambda_s(m)$ for every p. On the other hand, the cohomological assumption ensures that $\lambda_u(p, \gamma_s(p)) = \lambda_u(m)$. Taking $q = \gamma_s(p)$ we obtain

$$\lambda_u(p,q) = \lambda_u(m) \quad \text{and} \quad \lambda_s(p,q) = \lambda_s(m).$$

Hence, it follows from Lemma 6.5 that $m = \nu_{p,q}$. □

An analogous argument establishes the following result.

Lemma 6.8 *If $\lambda_u(m) \in I_u$ and ζ_s is cohomologous to a constant, then there exist $p, q \in \mathbb{R}$ such that $m = \nu_{p,q}$.*

Finally we consider the fourth case.

Lemma 6.9 *If $\lambda_u(m) \notin I_u$ and $\lambda_s(m) \notin I_s$, then:*

1. $\lambda_u(m) = \lambda_u^{\min}$ *and* $\lambda_s(m) = \lambda_s^{\max}$;
2. *there exists a measure $\nu \in \mathcal{M}_E$ such that*

$$\lambda_u(\nu) = \lambda_u(m), \quad \lambda_s(\nu) = \lambda_s(m) \quad \textit{and} \quad h_\nu(\Phi) = h_m(\Phi).$$

Proof of the lemma We first establish property 1. When $I_u = I_s = \varnothing$ (that is, when ζ_u and ζ_s are both cohomologous to constants) there is nothing to prove. Now let us assume that

$$I_u = \varnothing, \quad I_s \neq \varnothing \quad \text{and} \quad \lambda_s(m) = \lambda_s^{\min}. \tag{6.44}$$

Since $\nu_{0,0}$ is the measure of maximal entropy, we have $h(0,0) \geq h_m(\Phi)$. Hence, it follows from $\lambda_u(0,0) = \lambda_u^{\min}$, statement 2a in Lemma 6.1 and (6.19) that $\dim_H \nu_{0,0} > d(m)$. But this contradicts (6.35), and hence (6.44) cannot occur. Analogously, one can show that it is impossible to have $I_s = \varnothing$, $I_u \neq \varnothing$ and $\lambda_u(m) = \lambda_u^{\max}$.

To complete the proof of property 1, it remains to consider the case when $I_u \neq \varnothing$ and $I_s \neq \varnothing$. Then

$$\lambda_u(m) \in \partial I_u = \{\lambda_u^{\min}, \lambda_u^{\max}\} \quad \text{and} \quad \lambda_s(m) \in \partial I_s = \{\lambda_s^{\min}, \lambda_s^{\max}\}.$$

We first assume that

$$\lambda_u(m) = \lambda_u^{\max} \quad \text{and} \quad \lambda_s(m) = \lambda_s^{\min}. \tag{6.45}$$

Since $\nu_{0,0}$ is the measure of maximal entropy, we have $h(0,0) \geq h_m(\Phi)$. On the other hand, it follows from Lemma 6.1 that

$$\lambda_u(0,0) < \lambda_u(m) \quad \text{and} \quad \lambda_s(0,0) > \lambda_s(m).$$

By (6.19), we obtain $\dim_H \nu_{0,0} > d(m)$. But this contradicts (6.35), and hence (6.45) cannot occur. Now let us assume that

$$\lambda_u(m) = \lambda_u^{\min} \quad \text{and} \quad \lambda_s(m) = \lambda_s^{\min}. \tag{6.46}$$

We claim that

$$h(p,0) > h_m(\Phi) \tag{6.47}$$

for $p > 0$. Otherwise, if $h(p,0) \le h_m(\Phi)$ for some $p > 0$, then it would follow from Lemma 6.1 that

$$h(p,0) - p\lambda_u(p,0) < h_m(\Phi) - p\lambda_u(m).$$

But this is impossible, because $\nu_{p,0}$ is the equilibrium measure of $-p\zeta_u$. We also claim that

$$d_u(p,0) \ge h_m(\Phi)/\lambda_u(m) \tag{6.48}$$

for any sufficiently large p (see (6.27) for the definition of the function d_u). Otherwise, by Lemma 6.1, there would exist $p_0 \in \mathbb{R}$ and $\delta > 0$ such that

$$d_u(p,0) + \delta < h_m(\Phi)/\lambda_u(m)$$

for $p \ge p_0$. Then it would follow from (6.30) that $h_m(\Phi) > h(p,0)$ for any sufficiently large p. But this contradicts (6.47), and hence (6.48) holds for any sufficiently large p. By (6.46), (6.47) and (6.48), we obtain

$$\dim_H \nu_{p,0} = d_u(p,0) + d_s(p,0) \ge \frac{h_m(\Phi)}{\lambda_u(m)} - \frac{h(p,0)}{\lambda_s(p,0)} > d(m),$$

also for any sufficiently large p. This contradicts (6.35), and hence (6.46) cannot occur. Analogously, one can show that it is impossible to have

$$\lambda_u(m) = \lambda_u^{\max} \quad \text{and} \quad \lambda_s(m) = \lambda_s^{\max}.$$

This establishes property 1.

To establish property 2, consider an ergodic decomposition τ of the measure m (see Definition 4.9). Then

$$\lambda_u^{\min} = \lambda_u(m) = \int_{\mathcal{M}} \lambda_u(\nu)\,d\tau(\nu).$$

Since $\lambda_u(\nu) \ge \lambda_u^{\min}$ for $\nu \in \mathcal{M}$, there exists a set $A_1 \subset \mathcal{M}_E$ with $\tau(A_1) = 1$ such that $\lambda_u(\nu) = \lambda_u^{\min}$ for $\nu \in A_1$. Analogously, there exists a set $A_2 \subset \mathcal{M}_E$ with $\tau(A_2) = 1$ such that $\lambda_s(\nu) = \lambda_s^{\max}$ for $\nu \in A_2$. Hence, it follows from (6.19) and (6.35) that $h_\nu(\Phi) \le h_m(\Phi)$ for $\nu \in A_1 \cap A_2$. On the other hand, since

$$\tau(A_1 \cap A_2) = 1 \quad \text{and} \quad h_m(\Phi) = \int_{\mathcal{M}} h_\nu(\Phi)\,d\tau(\nu),$$

there exists a set $A \subset A_1 \cap A_2$ with $\tau(A) = 1$ such that $h_\nu(\Phi) = h_m(\Phi)$ for $\nu \in A$. This completes the proof of the lemma. □

By Lemmas 6.6, 6.7, 6.8 and 6.9, in each of the above four cases there exists a measure $\mu \in \mathcal{M}_E$ satisfying (6.37) (namely, the measure $\nu_{p,q}$ in the first three lemmas, and the measure ν in Lemma 6.9). This completes the proof of the theorem. □

Part III
Multifractal Analysis

This part is dedicated to the multifractal analysis of hyperbolic flows. In Chap. 7 we consider the simpler case of suspension flows over topological Markov chains. This allows us to present the main ideas without the additional technical complications that appear when one considers hyperbolic flows. We also show that for every Hölder continuous function noncohomologous to a constant the set of points without Birkhoff average has full topological entropy. In Chap. 8 we describe the multifractal analysis of hyperbolic flows. In the particular case of the entropy spectra, we show that the cohomology assumptions in the study of irregular sets are generically satisfied.

Chapter 7
Suspensions over Symbolic Dynamics

In this chapter we initiate the study of multifractal analysis for flows. This corresponds to giving a detailed description of the entropy of the level sets of the pointwise dimension for an invariant measure. We consider suspension flows over a topological Markov chain, which can be seen as a model for hyperbolic flows, although without certain additional technical complications. We refer to Chap. 8 for a multifractal analysis of hyperbolic flows. A nontrivial consequence of the results in this chapter is that for every Hölder continuous function that is not cohomologous to a constant, the set of points without Birkhoff average has full topological entropy. These results can essentially be proven in two different ways, either reducing the problem to the discrete-time dynamics on the base, or in an intrinsic manner, without leaving the context of flows. In order to include both approaches, in this chapter we consider the first approach, which consists of first reducing the problem to the dynamics on the base and then applying the existing results for discrete time. In Chap. 10 we obtain generalizations of the results in this chapter, using intrinsic arguments.

7.1 Pointwise Dimension

In this section we introduce the notion of pointwise dimension, in the general context of the BS-dimension. This is the local quantity that we will be considering in this chapter (we recall that any multifractal analysis consists of studying the complexity of the level sets of some local quantity).

Let $\Psi = \{\psi_t\}_{t\in\mathbb{R}}$ be a suspension flow in Y, over a homeomorphism $T\colon X \to X$ of the compact metric space X, and let μ be a T-invariant probability measure on X. We equip the space Y with the Bowen–Walters distance (see Sect. 2.2), and we consider the Φ-invariant probability measure ν induced by μ on Y.

Now we introduce the notion of pointwise dimension.

L. Barreira, *Dimension Theory of Hyperbolic Flows*,
Springer Monographs in Mathematics, DOI 10.1007/978-3-319-00548-5_7,

Definition 7.1 Given a continuous function $u\colon Y \to \mathbb{R}^+$, we define the *lower* and *upper u-pointwise dimensions* of ν at $x \in Y$ respectively by

$$\underline{d}_{\nu,u}(x) = \lim_{\varepsilon\to 0} \liminf_{t\to\infty} -\frac{\log \nu(B(x,t,\varepsilon))}{u(x,t,\varepsilon)} \tag{7.1}$$

and

$$\overline{d}_{\nu,u}(x) = \lim_{\varepsilon\to 0} \limsup_{t\to\infty} -\frac{\log \nu(B(x,t,\varepsilon))}{u(x,t,\varepsilon)}, \tag{7.2}$$

with $u(x,t,\varepsilon)$ as in (4.2).

For example, when Ψ is a suspension flow over a conformal expanding map T (see Sect. 8.1), for $u = \log \|dT\|$ the numbers $\underline{d}_{\nu,u}(x)$ and $\overline{d}_{\nu,u}(x)$ coincide respectively with the lower and upper pointwise dimensions of the measure ν, and for $u = 1$ these numbers coincide respectively with the lower and upper local entropies.

When ν is an ergodic Φ-invariant probability measure on Y, we have

$$\underline{d}_{\nu,u}(x) = \overline{d}_{\nu,u}(x) = \dim_u \nu = \frac{h_\nu(\Phi)}{\int_Y u\,d\nu}$$

for ν-almost every $x \in Y$. These identities can be obtained in a similar manner to that in the case of discrete time (see [3, Proposition 7.2.7]).

The following result shows that when T is a topological Markov chain the limits when $\varepsilon \to 0$ in (7.1) and (7.2) are not necessary.

Proposition 7.1 ([12]) *Let Ψ be a suspension flow over a topologically mixing two-sided topological Markov chain, let ν be an equilibrium measure for a Hölder continuous function g (with respect to Ψ), and let $u\colon Y \to \mathbb{R}^+$ be a Hölder continuous function. Then*

$$\underline{d}_{\nu,u}(y) = \liminf_{t\to\infty} -\frac{\log \nu(B(y,t,\varepsilon))}{\int_0^t u(\psi_\tau(y))\,d\tau} \tag{7.3}$$

and

$$\overline{d}_{\nu,u}(y) = \limsup_{t\to\infty} -\frac{\log \nu(B(y,t,\varepsilon))}{\int_0^t u(\psi_\tau(y))\,d\tau}$$

for every $y \in Y$ and any sufficiently small $\varepsilon > 0$.

Proof Given $\varepsilon > 0$, let

$$\delta(\varepsilon) = \sup\big\{|u(y_1) - u(y_2)| : d_Y(y_1, y_2) < \varepsilon\big\}.$$

Clearly, $\delta(\varepsilon) \to 0$ when $\varepsilon \to 0$. We have

$$\begin{aligned} 1 &\le \frac{u(y,t,\varepsilon)}{\int_0^t u(\psi_\tau(y))\,d\tau} \\ &\le \frac{\int_0^t [u(\psi_\tau(y)) + \delta(\varepsilon)]\,d\tau}{\int_0^t u(\psi_\tau(y))\,d\tau} \\ &\le 1 + \frac{\delta(\varepsilon)}{\inf u}, \end{aligned}$$

and thus,

$$\underline{d}_{\nu,u}(y) = \lim_{\varepsilon\to 0} \liminf_{t\to\infty} -\frac{\log \nu(B(y,t,\varepsilon))}{\int_0^t u(\psi_\tau(y))\,d\tau}. \tag{7.4}$$

For each $m \in \mathbb{N}$, let $\tau_m\colon X \to \mathbb{R}$ be the function in (2.13). Given $x \in X$, let $m = m(x,t) \in \mathbb{N}$ be the unique integer such that $\tau_{m-1}(x) \le t < \tau_m(x)$. By Proposition 2.3, there exists a constant $c \ge 1$ such that

$$B_X(x,m,\varepsilon) \times \left(s - \frac{\varepsilon}{c}, s + \frac{\varepsilon}{c}\right) \subset B(y,t,\varepsilon) \subset B_X(x,m-1,\varepsilon) \times (s - c\varepsilon, s + c\varepsilon) \tag{7.5}$$

for every $y = (x,s) \in Y$ and $t > 0$, and any sufficiently small $\varepsilon > 0$, where

$$B_X(x,m,\varepsilon) = \left\{ z \in X : d_X(T^k(z), T^k(x)) < \varepsilon \text{ for } k = 0,\dots,m \right\}. \tag{7.6}$$

By (4.8), this implies that

$$\liminf_{t\to\infty} -\frac{\log \nu(B(y,t,\varepsilon))}{\int_0^t u(\psi_\tau(y))\,d\tau} = \liminf_{t\to\infty} -\frac{\log \mu(B_X(x,m,\varepsilon))}{\int_0^t u(\psi_\tau(y))\,d\tau}. \tag{7.7}$$

On the other hand, by Proposition 2.2, the function I_g is Hölder continuous in X. Hence, since μ is an equilibrium measure for I_g it has the Gibbs property, that is, given $\varepsilon > 0$, there exists a $d \ge 1$ such that

$$d^{-1} \le \frac{\mu(B_X(x,m,\varepsilon))}{\exp\left(-mP_T(I_g) + \sum_{k=0}^{m-1} I_g(T^k(x))\right)} \le d \tag{7.8}$$

for every $x \in X$ and $m \in \mathbb{N}$. Thus, the limit in (7.7) is independent of ε, and it follows from (7.4) that identity (7.3) holds. A similar argument applies to $\overline{d}_{\nu,u}(y)$. □

7.2 Multifractal Analysis

In this section we present a multifractal analysis of the dimension spectrum for the u-pointwise dimension for suspension flows over a topological Markov chain.

Let Ψ be a suspension flow over T and let μ be a Ψ-invariant probability measure on Y. For each $\alpha \in \mathbb{R}$, let

$$K_\alpha = \{y \in Y : \underline{d}_{\nu,u}(y) = \overline{d}_{\nu,u}(y) = \alpha\}. \tag{7.9}$$

For $y \in K_\alpha$, the common value α of $\underline{d}_{\nu,u}(y)$ and $\overline{d}_{\nu,u}(y)$ is denoted by $d_{\nu,u}(y)$ and is called the *u-pointwise dimension of ν at the point y*.

Definition 7.2 The function $\mathcal{D}_u$ defined by

$$\mathcal{D}_u(\alpha) = \dim_u K_\alpha$$

is called the *u-dimension spectrum for the u-pointwise dimensions* (with respect to the measure ν).

For example, if $u = 1$, then $\mathcal{D}_u(\alpha) = h(\Phi|K_\alpha)$.

Now let $g: Y \to \mathbb{R}$ be a Hölder continuous function. For each $q \in \mathbb{R}$, we define a function $g_q: Y \to \mathbb{R}$ by

$$g_q = -T_u(q)u + qg,$$

where $T_u(q)$ is the unique real number such that $P_\Psi(g_q) = 0$. We denote respectively by ν_q and m_u the equilibrium measures for g_q and $-\dim_u Y \cdot u$ (with respect to Ψ). The conditions in Theorem 7.1 below ensure that $T_u(q)$, ν_q and m_u are uniquely defined.

The following is a multifractal analysis of the spectrum $\mathcal{D}_u$ for suspension flows over topological Markov chains.

Theorem 7.1 ([12]) *Let Ψ be a suspension flow over a topologically mixing two-sided topological Markov chain, let $u: Y \to \mathbb{R}^+$ be a Hölder continuous function, and let ν be an equilibrium measure for a Hölder continuous function g such that $P_\Psi(g) = 0$. Then the following properties hold:*

1. *the function T_u is analytic, $T_u'(q) \le 0$ and $T_u''(q) \ge 0$ for every $q \in \mathbb{R}$, $T_u(0) = \dim_u Y$ and $T_u(1) = 0$;*
2. *the domain of $\mathcal{D}_u$ is a closed interval in $[0, \infty)$ and coincides with the range of the function $\alpha_u = -T_u'$;*
3. *for each $q \in \mathbb{R}$ we have $\nu_q(K_{\alpha_u(q)}) = 1$,*

$$\mathcal{D}_u(\alpha_u(q)) = T_u(q) + q\alpha_u(q) = \dim_u \nu_q,$$

$$d_{\nu_q,u}(x) = T_u(q) + q\alpha_u(q)$$

for ν_q-almost every $x \in K_{\alpha_u(q)}$, and

$$\overline{d}_{\nu_q,u}(x) \le T_u(q) + q\alpha_u(q)$$

for every $x \in K_{\alpha_u(q)}$;

4. *if $\nu \neq m_u$, then $\mathcal{D}_u$ and T_u are analytic strictly convex functions.*

Proof Again, the idea of the proof is to reduce the problem to the case of maps. We first express the pointwise dimension in terms of the dynamics in the base.

Lemma 7.1 *If $y = (x, s) \in Y$, then*

$$\underline{d}_{\nu,u}(y) = \liminf_{m\to\infty} -\frac{\sum_{i=0}^m I_g(T^i(x))}{\sum_{i=0}^m I_u(T^i(x))}$$

and

$$\overline{d}_{\nu,u}(y) = \limsup_{m\to\infty} -\frac{\sum_{i=0}^m I_g(T^i(x))}{\sum_{i=0}^m I_u(T^i(x))}.$$

Proof of the lemma Let $\tau_m : Y \to \mathbb{R}$ be the function in (2.13). Given $t > 0$, let $m \in \mathbb{N}$ be the unique integer such that $\tau_m(x) \le t < \tau_{m+1}(x)$, and write $t = \tau_m(x) + \kappa$ with $\kappa \in (\inf \tau, \sup \tau)$. Proceeding as in the proof of Theorem 2.3, we obtain

$$\left| \frac{1}{t} \int_0^t u(\psi_\tau(y))\, d\tau - \frac{1}{\tau_m(y)} \sum_{i=0}^{m-1} I_u(T^i(y)) \right| \to 0 \tag{7.10}$$

when $t \to \infty$. Now let $B_X(x, m, \varepsilon)$ be the Bowen ball in (7.6). By (7.5), we have

$$\left| \frac{-\log \nu(B(y, t, \varepsilon))}{t} + \frac{\log \mu(B_X(x, m, \varepsilon))}{\tau_m(x)} \right| \to 0 \tag{7.11}$$

when $t \to \infty$. Moreover, $T^i(x, s) = T^i(x, 0)$ for every $i \in \mathbb{N}$, and hence,

$$\sum_{i=0}^{m-1} I_u(T^i(y)) = \sum_{i=0}^{m-1} I_u(T^i(x)).$$

Now let

$$A = \frac{-\log \nu(B(y, t, \varepsilon))}{\int_0^t u(\psi_\tau(y))\, d\tau} + \frac{\log \mu(B_X(x, m, \varepsilon))}{\sum_{i=0}^{m-1} I_u(T^i(x))}.$$

Since $0 < \inf u \le \sup u < \infty$, it follows from (7.10) and (7.11) that

$$A = \left(\frac{-\log \mu(B_X(x, m, \varepsilon))}{\tau_m(x)} + o(t) \right) \frac{t}{\int_0^t u(\psi_\tau(y))\, d\tau}$$

$$+ \frac{\log \mu(B_X(x, m, \varepsilon))}{\tau_m(x)} \left(\frac{t}{\int_0^t u(\psi_\tau(y))\, d\tau} + o(t) \right),$$

and hence,

$$|A| \le \left(\frac{1}{\inf u} + \frac{h_\mu(T)}{\inf \tau} \right) o(t).$$

This completes the proof of the lemma. □

We also express the BS-dimension in terms of a Carathéodory characteristic in the base. Given a set $Z \subset X$ and $\beta \in \mathbb{R}$, let

$$N_\beta(Z) = \lim_{\ell\to\infty} \inf_{\Gamma} \sum_{C\in\Gamma} \exp\left(-\beta \sup\left\{\sum_{i=0}^{m(C)-1} I_u(T^i(x)) : x \in C\right\}\right), \tag{7.12}$$

where the infimum is taken over all finite or countable covers Γ of Z by cylinder sets

$$C_{i_{-n}\cdots i_m} = \{(\cdots j_0 \cdots) : j_k = i_k \text{ for } -n \le k \le m\}, \quad \text{with} \quad m, n > \ell. \tag{7.13}$$

Lemma 7.2 *If the set $Z \subset X$ is T-invariant, then*

$$\dim_u\{(x,s) \in Y : x \in Z \text{ and } s \in [0, \tau(x)]\} = \inf\{\beta \in \mathbb{R} : N_\beta(Z) = 0\}.$$

Proof of the lemma Using the same notation as in the proof of Lemma 7.1, we obtain the inequality

$$\left| \int_0^t u(\psi_\tau(x))\, d\tau - \sum_{i=0}^{m-1} I_u(T^i(x)) \right| \le \kappa \sup u,$$

which yields the desired result. □

The above lemmas allow us to reduce the study of the spectrum $\mathcal{D}_u$ to the study of corresponding properties of the dynamics in the base. Namely, by Lemma 7.1, we have

$$K_\alpha = \{(x,s) \in Y : x \in Z_\alpha \text{ and } s \in [0, \tau(x)]\},$$

where

$$Z_\alpha = \left\{ x \in X : \lim_{m\to\infty} -\frac{\sum_{i=0}^{m-1} I_g(T^i(x))}{\sum_{i=0}^{m-1} I_u(T^i(x))} = \alpha \right\},$$

and it follows from Lemma 7.2 that

$$\mathcal{D}_u(\alpha) = \inf\{\beta \in \mathbb{R} : N_\beta(Z_\alpha) = 0\}.$$

In other words, the u-dimension spectrum $\mathcal{D}_u$ for the u-pointwise dimensions (with respect to the measure ν) coincides with the I_g-dimension spectrum studied by Barreira and Schmeling in [17] (in the case of discrete time). Hence, the desired result follows readily from Theorem 6.6 in that paper (see [3] for a detailed discussion). □

Theorem 7.1 is a continuous-time version of Theorem 6.6 in [17], which follows from work of Pesin and Weiss [83] and Schmeling [97].

Taking $u = 1$ in Theorem 7.1, we obtain a multifractal analysis of the spectrum

$$\mathcal{E}(\alpha) = h\big(\Psi|\{y \in Y : h_\nu(y) = \alpha\}\big),$$

where

$$h_\nu(y) = \lim_{t\to\infty} -\frac{\log \nu(B(y,t,\varepsilon))}{t} = \lim_{t\to\infty} \frac{1}{t}\int_0^t g(\psi_\tau(y))\,d\tau \qquad (7.14)$$

(see Theorem 7.3). The function $\mathcal{E}$ is called the *entropy spectrum for the local entropies* (with respect to the measure ν), and coincides with the entropy spectrum for the Birkhoff averages of g (see Sect. 7.4).

7.3 Irregular Sets

In this section we consider the complement of the sets K_α in (7.9), that is, the *irregular set* $Z = Y \setminus \bigcup_{\alpha\in\mathbb{R}} K_\alpha$. Even though Z has zero measure with respect to any Φ-invariant probability measure it also has full BS-dimension.

Let $\Psi = \{\psi_t\}_{t\in\mathbb{R}}$ be a continuous flow in Y.

Definition 7.3 Given continuous functions $g_1, \ldots, g_k \colon Y \to \mathbb{R}$ and $u \colon Y \to \mathbb{R}^+$, we consider the *irregular set*

$$\mathcal{F}_u(g_1, \ldots, g_k) = \bigcap_{j=1}^k \left\{ y \in Y : \liminf_{t\to\infty} \frac{\int_0^t g_j(\psi_s(y))\,ds}{\int_0^t u(\psi_s(y))\,ds} < \limsup_{t\to\infty} \frac{\int_0^t g_j(\psi_s(y))\,ds}{\int_0^t u(\psi_s(y))\,ds} \right\}.$$

We have

$$\mathcal{F}_u(g_1, \ldots, g_k) = \big\{(x,s) : x \in \mathcal{C}_u(g_1, \ldots, g_k) \text{ and } s \in [0, \tau(x)]\big\},$$

where

$$\mathcal{C}_u(g_1, \ldots, g_k) = \bigcap_{j=1}^k \left\{ x \in X : \liminf_{m\to\infty} \frac{\sum_{i=0}^m I_{g_j}(T^i(x))}{\sum_{i=0}^m I_u(T^i(x))} < \limsup_{m\to\infty} \frac{\sum_{i=0}^m I_{g_j}(T^i(x))}{\sum_{i=0}^m I_u(T^i(x))} \right\}.$$

This is a consequence of the following result.

Proposition 7.2 *Let Ψ be a suspension flow over a map $T \colon X \to X$ and let $a, b \colon Y \to \mathbb{R}$ be continuous functions with $b > 0$. If $x \in X$ and $s \in [0, \tau(x)]$, then*

$$\liminf_{t\to\infty} \frac{\int_0^t a(\psi_\tau(x,s))\,d\tau}{\int_0^t b(\psi_\tau(x,s))\,d\tau} = \liminf_{m\to\infty} \frac{\sum_{i=0}^m I_a(T^i(x))}{\sum_{i=0}^m I_b(T^i(x))}$$

and

$$\limsup_{t\to\infty}\frac{\int_0^t a(\psi_\tau(x,s))\,d\tau}{\int_0^t b(\psi_\tau(x,s))\,d\tau}=\limsup_{m\to\infty}\frac{\sum_{i=0}^m I_a(T^i(x))}{\sum_{i=0}^m I_b(T^i(x))}.$$

Proof The argument is a modification of the proof of Theorem 2.3. Given $m\in\mathbb{N}$, we consider the function $\tau_m\colon Y\to\mathbb{R}$ in (2.13). For each $t>0$, there exists a unique integer $m\in\mathbb{N}$ such that $\tau_m(x)\le t<\tau_{m+1}(x)$. Writing $t=\tau_m(x)+\kappa$ with $\kappa\in(\inf\tau,\sup\tau)$, we obtain

$$\begin{aligned}
&\left|\frac{\int_0^t a(\psi_s(x))\,ds}{\int_0^t b(\psi_s(x))\,ds}-\frac{\int_0^{\tau_m(x)} a(\psi_s(x))\,ds}{\int_0^{\tau_m(x)} b(\psi_s(x))\,ds}\right|\\
&=\left|\frac{\int_{\tau_m(x)}^t a(\psi_s(x))\,ds\int_0^{\tau_m(x)} b(\psi_s(x))\,ds-\int_0^{\tau_m(x)} a(\psi_s(x))\,ds\int_{\tau_m(x)}^t b(\psi_s(x))\,ds}{\int_0^t b(\psi_s(x))\,ds\int_0^{\tau_m(x)} b(\psi_s(x))\,ds}\right|\\
&\le\frac{\kappa\sup|a|\cdot\tau_m(x)\sup b+\tau_m(x)\sup|a|\cdot\kappa\sup b}{\tau_m(x)\sup b\cdot\tau_m(x)\sup b}\\
&=\frac{2\kappa\sup|a|}{\tau_m(x)\sup b}.
\end{aligned}$$

Letting $t\to\infty$, we have $m\to\infty$ and $\tau_m(x)\to\infty$. Hence, it follows from (2.14) that

$$\left|\frac{\int_0^t a(\psi_s(x))\,ds}{\int_0^t b(\psi_s(x))\,ds}-\frac{\sum_{i=0}^{m-1} I_a(T^i(x))}{\sum_{i=0}^{m-1} I_b(T^i(x))}\right|\to 0$$

when $t\to\infty$. This yields the desired result. □

In particular, Proposition 7.2 allows us to reduce the study of irregular sets to the study of corresponding sets in the base.

The following result gives a necessary and sufficient condition so that the irregular set $\mathfrak{F}_u(g_1,\dots,g_k)$ has full BS-dimension.

Theorem 7.2 ([12]) *Let Ψ be a suspension flow over a topologically mixing two-sided topological Markov chain and let $g_1,\dots,g_k,u\colon Y\to\mathbb{R}$ be Hölder continuous functions with $u>0$. Then the following properties are equivalent:*

1. *g_j is not Ψ-cohomologous to a multiple of u in Y, for $j=1,\dots,k$;*
2. $\dim_u\mathfrak{F}_u(g_1,\dots,g_k)=\dim_u Y$.

Proof In a similar manner to that in the proof of Theorem 7.1, we first reduce the problem to the case of maps. By Lemma 7.2, we have

$$\dim_u Y=\inf\{\beta\in\mathbb{R}:N_\beta(X)=0\}\tag{7.15}$$

and

$$\dim_u \mathcal{F}_u(g_1,\ldots,g_k) = \inf\{\beta \in \mathbb{R} : N_\beta(\mathcal{C}_u(g_1,\ldots,g_k)) = 0\}, \tag{7.16}$$

with $N_\beta(Z)$ as in (7.12). We note that the set $\mathcal{C}_u(g_1,\ldots,g_k)$ is defined entirely in terms of the map T and the functions I_u and I_{g_j} for $j=1,\ldots,k$. On the other hand, by Theorem 2.1, the function g_j is Ψ-cohomologous to a multiple of u in Y if and only if I_{g_j} is T-cohomologous to a multiple of I_u in X, and hence, if and only if I_{g_j} is T-cohomologous to $I_{\alpha_j u} = \alpha_j I_u$ in X, where α_j is the unique real number such that $P_T(I_{g_j}) = P_T(\alpha_j I_u)$. These are precisely the cohomology assumptions in Theorem 7.1 in [17] in the case of discrete time (see [3] for a detailed discussion), which tell us that g_j is not Ψ-cohomologous to a multiple of u in Y, for $j=1,\ldots,k$, if and only if the right-hand sides of (7.15) and (7.16) are equal. This yields the desired result. □

7.4 Entropy Spectra

This section considers the particular case of entropy spectra. As a consequence of the results in the former sections, we obtain a multifractal analysis of these spectra and we study the corresponding irregular sets.

Let Ψ be a suspension flow over a map $T\colon X \to X$ and let $g\colon Y \to \mathbb{R}$ be a continuous function. For each $\alpha \in \mathbb{R}$, let

$$\mathcal{E}(\alpha) = h(\Psi|K_\alpha),$$

where

$$K_\alpha = \left\{x \in Y : \lim_{t\to\infty} \frac{1}{t}\int_0^t g(\psi_\tau(x))\,d\tau = \alpha\right\}.$$

The topological entropy is computed with respect to the Bowen Walters distance in Y. The function $\mathcal{E}$ is called the *entropy spectrum for the Birkhoff averages of g*. For each $q \in \mathbb{R}$, let ν_q be the equilibrium measure for qg and write

$$T(q) = P_\Psi(qg).$$

The following result is a multifractal analysis of the spectrum $\mathcal{E}$.

Theorem 7.3 ([12]) *Let Ψ be a suspension flow over a topologically mixing two-sided topological Markov chain and let $g\colon Y \to \mathbb{R}$ be a Hölder continuous function with $P_\Psi(g) = 0$. Then the following properties hold*:

1. *the domain of $\mathcal{E}$ is a closed interval in $[0,\infty)$ coinciding with the range of the function $\alpha = -T'$, and for each $q \in \mathbb{R}$ we have*

$$\mathcal{E}(\alpha(q)) = T(q) + q\alpha(q) = h_{\nu_q}(\Psi);$$

2. *if g is not Ψ-cohomologous to a constant in Y, then $\mathcal{E}$ and T are analytic strictly convex functions.*

Proof In view of (7.14), the result follows from Theorem 7.1 taking $u = 1$. □

Now we consider the irregular sets. More precisely, given a continuous function $g\colon Y \to \mathbb{R}$, let

$$\mathcal{B}(g) = \left\{ y \in Y : \liminf_{t\to\infty} \frac{1}{t} \int_0^t g(\psi_\tau(y))\, d\tau < \limsup_{t\to\infty} \frac{1}{t} \int_0^t g(\psi_\tau(y))\, d\tau \right\}.$$

It follows from Theorem 2.3 that

$$\mathcal{B}(g) = \big\{ (x, s) \in Y : x \in \mathcal{C} \text{ and } s \in [0, \tau(x)] \big\},$$

where

$$\mathcal{C} = \left\{ x \in X : \liminf_{m\to\infty} \frac{\sum_{i=0}^m I_g(T^i(x))}{\sum_{i=0}^m \tau(T^i(x))} < \limsup_{m\to\infty} \frac{\sum_{i=0}^m I_g(T^i(x))}{\sum_{i=0}^m \tau(T^i(x))} \right\}.$$

The following result gives a necessary and sufficient condition so that the irregular sets have full topological entropy.

Theorem 7.4 ([12]) *Let Ψ be a suspension flow over a topologically mixing two-sided topological Markov chain and let $g_j\colon Y \to \mathbb{R}$ be Hölder continuous functions for $j = 1, \ldots, k$. Then the following properties are equivalent*:

1. *g_j is not Ψ-cohomologous to a constant in Y, for $j = 1, \ldots, k$;*
2. *$h(\Psi | \bigcap_{j=1}^k \mathcal{B}(g_j)) = h(\Psi)$.*

Proof The result follows from Theorem 7.2 taking $k = 1$, $g_1 = g$ and $u = 1$. □

Chapter 8
Multifractal Analysis of Hyperbolic Flows

In this chapter we continue the study of multifractal analysis for flows. The emphasis is now on dimension spectra of hyperbolic flows. We first consider the somewhat simpler case of suspension semiflows over expanding maps. It is presented mainly as a motivation for the case of hyperbolic sets for conformal flows, without the additional complication of simultaneously having contraction and expansion. In the case of entropy spectra for hyperbolic flows, we show that the cohomology assumptions required in the study of irregular sets are generically satisfied.

8.1 Suspensions over Expanding Maps

In this section we consider suspension semiflows over conformal expanding maps and we obtain a multifractal analysis of the dimension spectra of Gibbs measures. This can be seen as a simplified version of the multifractal analysis of the dimension spectra of Gibbs measures on locally maximal hyperbolic sets, without simultaneously having expansion and contraction.

Let $f\colon M \to M$ be a C^1 map of a smooth manifold M and let $\Lambda \subset M$ be a compact f-invariant set such that f is expanding on Λ. This means that there exist constants $c > 0$ and $\beta > 1$ such that

$$\|d_x f^n v\| \ge c\beta^n \|v\|$$

for all $x \in \Lambda$, $v \in T_x M$ and $n \in \mathbb{N}$. The set Λ is said to be a *repeller* of f. We note that any repeller of f is also a repeller of f^n for each $n \in \mathbb{N}$. Thus, passing eventually to a power of f, without loss of generality one can take $c = 1$. For simplicity of the exposition we always make this assumption.

We also introduce the notion of conformality.

Definition 8.1 The map f is said to be *conformal* on Λ if $d_x f$ is a multiple of an isometry for every $x \in \Lambda$.

L. Barreira, *Dimension Theory of Hyperbolic Flows*,
Springer Monographs in Mathematics, DOI 10.1007/978-3-319-00548-5_8,

Any repeller has Markov partitions of arbitrarily small diameter. Each Markov partition has associated a one-sided topological Markov chain $\sigma: X \to X$ and a coding map $\pi: X \to \Lambda$ for the repeller. The map π is onto, finite-to-one, and satisfies $f \circ \pi = \pi \circ \sigma$. We refer to [5] for details.

Consider a Markov partition of Λ and the associated coding map $\pi: X \to \Lambda$. Let Ψ be the suspension semiflow over the one-sided topological Markov chain $\sigma: X \to X$. We introduce a distance d_X in X (inducing the usual topology) with the property that for a repeller Λ of a conformal map the coding map $\pi: (X, d_X) \to \Lambda$ is locally Lipschitz. Let $u: X \to \mathbb{R}^+$ be the continuous function

$$u(x) = \log\|d_{\pi(x)} f\|. \tag{8.1}$$

The distance d_X is defined by

$$d_X\big((i_0 \cdots), (j_0 \cdots)\big) = |i_0 - j_0| + \exp\big(-u(C_{i_0 \cdots i_n})\big),$$

where

$$n = \max\big\{m \in \mathbb{N} : i_k = j_k \text{ for } k \le m\big\}$$

and

$$u(C_{i_0 \cdots i_n}) = \sup\left\{\sum_{k=0}^{n} u(\sigma^k(x)) : x \in C_{i_0 \cdots i_n}\right\}.$$

The set Y is now equipped with the corresponding Bowen–Walters distance. It follows from work of Schmeling [98] that if f is a $C^{1+\delta}$ expanding map that is conformal on Λ, then

$$\dim_H Z = 1 + \dim_u \pi^{-1} Z$$

for any Ψ-invariant set $Z \subset \Lambda$, with u as in (8.1). We note that here $\dim_u$ is the BS-dimension introduced by Barreira and Schmeling in [17] (for discrete time) and not the corresponding notion for continuous time described in Sect. 4.2.

Now we introduce the dimension spectrum. Let ν be a Ψ-invariant probability measure on Y. For each $\alpha \in \mathbb{R}$, let

$$K_\alpha = \left\{y \in Y : \lim_{r \to 0} \frac{\log \nu(B(y, r))}{\log r} = \alpha\right\},$$

where $B(y, r) \subset Y$ is the Bowen–Walters ball of radius r centered at $y \in Y$.

Definition 8.2 The function

$$\mathcal{D}(\alpha) = \dim_H K_\alpha$$

is called the *dimension spectrum for the pointwise dimensions* (with respect to the measure ν).

Let also μ be the measure on X associated to ν as in (4.8). By Proposition 2.3, there exists a $c \geq 1$ such that

$$B_X(x, r/c) \times (s - r/c, s + r/c) \subset B(y, r) \subset B_X(x, cr) \times (s - cr, s + cr)$$

for every $y = (x, s) \in Y$ and any sufficiently small r (taking the distance d_X in X). Therefore,

$$K_\alpha = \left\{ (x, s) \in Y : \lim_{r \to 0} \frac{\log \mu(B_X(x, r))}{\log r} = \alpha - 1 \right\}.$$

Since each set K_α is Ψ-invariant, we obtain

$$\mathcal{D}(\alpha) = 1 + \dim_u \left\{ x \in X : \lim_{r \to 0} \frac{\log \mu(B_X(x, r))}{\log r} = \alpha - 1 \right\}, \tag{8.2}$$

with u as in (8.1).

Now let $g\colon Y \to \mathbb{R}$ be a Hölder continuous function. For each $q \in \mathbb{R}$, we define a function $g_q\colon Y \to \mathbb{R}$ by

$$g_q = -T_u(q)u + qg,$$

where $T_u(q)$ is the unique real number such that $P_\Psi(g_q) = 0$. We denote respectively by ν_q and m_u the equilibrium measures for g_q and $-\dim_u Y \cdot u$ (with respect to Ψ).

Proceeding in a similar manner to that in Sect. 7.1 one can obtain a multifractal analysis of the spectrum $\mathcal{D}$. Using the same notation as in Sect. 7.1, the following result is a simple consequence of Theorem 7.1 and the above discussion, together with appropriate versions of Propositions 2.1 and 2.3 for locally invertible maps.

Theorem 8.1 ([12]) *Let Λ be a repeller of a $C^{1+\delta}$ map that is conformal and topologically mixing on Λ and let Ψ be the suspension semiflow over the one-sided topological Markov chain associated to some Markov partition of Λ. If ν is an equilibrium measure for a Hölder continuous function g (with respect to Ψ), then the following properties hold:*

1. *for ν-almost every $y \in Y$ we have*

$$\lim_{r \to 0} \frac{\log \nu(B(y, r))}{\log r} = 1 + \frac{h_\mu(f)}{\int_X u \, d\mu};$$

2. *the domain of $\mathcal{D}$ is a closed interval in $[0, \infty)$ and coincides with the range of the function $\alpha = -T'$, where $T(q) = T_u(q) - q + 1$;*
3. *for each $q \in \mathbb{R}$ we have $\nu_q(K_{\alpha(q)}) = 1$,*

$$\mathcal{D}(\alpha(q)) = T(q) + q\alpha(q) = \dim_H \nu_q, \tag{8.3}$$

$$\lim_{r \to 0} \frac{\log \nu_q(B(y, r))}{\log r} = T(q) + q\alpha(q)$$

for ν_q-almost every $x \in K_{\alpha(q)}$, and

$$\limsup_{r\to 0} \frac{\log \nu_q(B(y,r))}{\log r} \le T(q) + q\alpha(q)$$

for every $x \in K_{\alpha(q)}$;

4. *if $\nu \ne m_u$, then $\mathcal{D}$ and T are analytic strictly convex functions.*

Let $\alpha_u(q) = -T_u'(q)$. By (8.2), we have

$$\begin{aligned} T(q) + q\alpha(q) &= T_u(q) - q + 1 + q(-T_u'(q) + 1) \\ &= 1 + T_u(q) + q\alpha_u(q) \\ &= 1 + \dim_u \left\{ x \in X : \lim_{r\to 0} \frac{\log \mu(B_X(x,r))}{\log r} = \alpha_u(q) \right\} \\ &= \mathcal{D}(\alpha_u(q) + 1) = \mathcal{D}(\alpha(q)), \end{aligned}$$

which establishes the first equality in (8.3).

In a similar manner, the following result follows easily from an appropriate version of Theorem 7.2 for suspension semiflows over one-sided topological Markov chains.

Theorem 8.2 *Under the hypotheses of Theorem* 8.1, *we have $\nu \ne m_u$ if and only if*

$$\dim_H \left\{ y \in Y : \liminf_{r\to 0} \frac{\log \nu(B(y,r))}{\log r} < \limsup_{r\to 0} \frac{\log \nu(B(y,r))}{\log r} \right\} = \dim_H Y.$$

8.2 Dimension Spectra of Hyperbolic Flows

In this section we obtain a multifractal analysis of the dimension spectrum for the pointwise dimensions of a Gibbs measure (on a locally maximal hyperbolic set for a conformal flow). This can be seen as an elaborate version of Theorem 8.1, in which case only expansion is present.

Let Φ be a C^1 flow with a locally maximal hyperbolic set Λ. Given a Markov system, we consider the associated two-sided topological Markov chain $\sigma : X \to X$ and the coding map $\pi : X \to \Lambda$ (see Sect. 3.2).

Now let $\beta_s, \beta_u : X \to \mathbb{R}^+$ be Hölder continuous functions. For each cylinder set $C_{i_{-n}\cdots i_m}$ in (7.13), we define

$$\beta_s(C_{i_{-n}\cdots i_m}) = \sup\left\{ \sum_{k=0}^{m} \beta_s(\sigma^k(x)) : x \in C_{i_{-n}\cdots i_m} \right\}$$

and

$$\beta_u(C_{i_{-n}\cdots i_m}) = \sup\left\{\sum_{k=0}^{n} \beta_u(\sigma^{-k}(x)) : x \in C_{i_{-n}\cdots i_m}\right\}.$$

For each set $Z \subset X$ and $\alpha \in \mathbb{R}$, let

$$M(Z,\alpha) = \lim_{\ell\to 0} \inf_{\Gamma} \sum_{C\in\Gamma} \exp\bigl(-\alpha\beta_s(C) - \alpha\beta_u(C)\bigr),$$

where the infimum is taken over all finite or countable covers Γ of Z by cylinder sets $C_{i_{-n}\cdots i_m}$ with $m, n > \ell$.

Definition 8.3 The (β_s, β_u)-*dimension* of the set Z is defined by

$$\dim_{\beta_s,\beta_u} Z = \inf\bigl\{\alpha \in \mathbb{R} : M(Z,\alpha) = 0\bigr\}.$$

Now let Φ be a $C^{1+\delta}$ flow with a locally maximal hyperbolic set Λ such that $\Phi|\Lambda$ is conformal. We continue to consider a Markov system for Φ on Λ and the associated symbolic dynamics. We define functions $\beta_s, \beta_u \colon X \to \mathbb{R}$ by

$$\beta_s = -I_{\zeta_s} \circ \pi \quad \text{and} \quad \beta_u = I_{\zeta_u} \circ \pi, \tag{8.4}$$

with ζ_s and ζ_u as in (5.2) and (5.3). One can easily verify that

$$\beta_s(x) = -\log\|d_{\pi(x)}\varphi_{\tau(\pi(x))}|E^s(\pi(x))\|$$

and

$$\beta_u(x) = \log\|d_{\pi(x)}\varphi_{\tau(\pi(x))}|E^u(\pi(x))\|.$$

Without loss of generality, one can always assume that β_s and β_u are positive functions (simply consider an adapted metric). Since Φ is conformal on Λ, we have

$$\sum_{k=0}^{n-1} \beta_s(\sigma^k(x)) = -\log\|d_{\pi(x)}\varphi_{\tau_n(\pi(x))}|E^s(\pi(x))\|$$

and

$$\sum_{k=0}^{n-1} \beta_u(\sigma^{-k}(x)) = \log\|d_{\pi(x)}\varphi_{-\tau_n(\pi(x))}|E^u(\pi(x))\|,$$

where

$$\tau_n(\pi(x)) = \sum_{k=0}^{n-1} \tau(\pi(\sigma^k(x))).$$

It follows from work of Schmeling [98] that

$$\dim_H Z = 1 + \dim_{\beta_s,\beta_u} \pi^{-1} Z$$

for any Ψ-invariant set $Z \subset \Lambda$.

When X is equipped with the distance d in (3.10), in general the map π is only Hölder continuous. We introduce a new distance d_X in X (inducing the same topology as d) such that for flows that are conformal on Λ the map $\pi\colon (X, d_X) \to \Lambda$ is locally Lipschitz. The new distance d_X is defined by

$$d_X\big((\cdots i_0 \cdots), (\cdots j_0 \cdots)\big) = |i_0 - j_0| + \exp\big(-\beta_s(C_{i_{-n_u}\cdots i_{n_s}})\big) + \exp\big(-\beta_u(C_{i_{-n_u}\cdots i_{n_s}})\big),$$

where

$$n_s = \max\{n \in \mathbb{N} : i_k = j_k \text{ for } k \le n\}$$

and

$$n_u = \max\{n \in \mathbb{N} : i_k = j_k \text{ for } k \ge -n\}.$$

Since

$$\operatorname{diam}_{d_X} C = \beta_s(C) + \beta_u(C)$$

for any cylinder set C, the (β_s, β_u)-dimension of a set $Z \subset X$ coincides with its Hausdorff dimension with respect to the distance d_X. This distance induces a new Bowen–Walters distance in Y.

Now we introduce the dimension spectrum. We continue to consider a locally maximal hyperbolic set for a C^1 flow Φ. Let also ν be a Φ-invariant probability measure on Λ. For each $\alpha \in \mathbb{R}$, let

$$K_\alpha = \left\{ y \in \Lambda : \lim_{r \to 0} \frac{\log \nu(B(y,r))}{\log r} = \alpha \right\}.$$

Definition 8.4 The *dimension spectrum for the pointwise dimensions* (with respect to the measure ν) is defined by

$$\mathcal{D}(\alpha) = \dim_H K_\alpha.$$

In a similar manner to that in (8.2), if Φ is conformal on Λ, then

$$\mathcal{D}(\alpha) = 1 + \dim_{\beta_s, \beta_u} \left\{ x \in X : \lim_{r \to 0} \frac{\log \mu(B_X(x,r))}{\log r} = \alpha - 1 \right\},$$

with β_s and β_u as in (8.4).

Given a continuous function $g\colon \Lambda \to \mathbb{R}$, for each $q \in \mathbb{R}$ let $T_s(q)$ and $T_u(q)$ be the unique real numbers such that

$$P_{\Phi|\Lambda}\big(T_s(q)\zeta_s + qg\big) = P_{\Phi|\Lambda}\big(-T_u(q)\zeta_u + qg\big) = 0,$$

or equivalently,

$$P_\sigma\big(-T_s(q)\beta_s + qI_g \circ \pi\big) = P_\sigma\big(-T_u(q)\beta_u + qI_g \circ \pi\big) = 0. \tag{8.5}$$

We write

$$T(q) = T_s(q) + T_u(q) - q + 1. \tag{8.6}$$

The following result is due to Pesin and Sadovskaya [82]. It is a multifractal analysis of the dimension spectrum for the pointwise dimensions of a Gibbs measure on a locally maximal hyperbolic set.

Theorem 8.3 *Let Φ be a $C^{1+\delta}$ flow with a locally maximal hyperbolic set Λ such that $\Phi|\Lambda$ is conformal and topologically mixing and let ν be an equilibrium measure for a Hölder continuous function g (with respect to Φ) such that $P_{\Phi|\Lambda}(g) = 0$. Then the following properties hold*:

1. *for ν-almost every $y \in \Lambda$ we have*

$$\lim_{r\to 0} \frac{\log \nu(B(y,r))}{\log r} = h_\mu(\Phi)\left(\frac{1}{\int_\Lambda \zeta_u\,d\mu} - \frac{1}{\int_\Lambda \zeta_s\,d\mu}\right) + 1;$$

2. *if $\alpha = -T'$, then*

$$\mathcal{D}(\alpha(q)) = T(q) + q\alpha(q) \quad \textit{for} \quad q \in \mathbb{R}.$$

Proof Since the measure ν is ergodic, the first property is an immediate consequence of Theorem 6.2.

To establish the second property, we proceed in a similar manner to that in the proof of Theorem 5.1. Namely, let $R_1, \ldots, R_k$ be a Markov system for Φ on Λ. We also consider the function τ in (3.4) and the map T in (3.5), where $Z = \bigcup_{i=1}^k R_i$. Let S be the invertible map $T|Z\colon Z \to Z$ and let A be the transition matrix obtained from the Markov system as in (3.8).

We recall the projections π_+ and π_- in (3.13) and (3.14). It follows from a construction described by Bowen in [28] (see Proposition 4.2.11 in [3]) that there exist functions $\psi^s, d^s\colon \Sigma_A^- \to \mathbb{R}$ and $\psi^u, d^u\colon \Sigma_A^+ \to \mathbb{R}$ such that:

1. $I_g \circ \pi$, $\psi^s \circ \pi_-$ and $\psi^u \circ \pi_+$ are cohomologous;
2. $\log \|dS^{-1}|E^s\| \circ \pi$ and $d^s \circ \pi_-$ are cohomologous;
3. $\log \|dT|E^u\| \circ \pi$ and $d^u \circ \pi_+$ are cohomologous.

Given $\omega \in \Sigma_A$ and $r \in (0,1)$, let $n = n(\omega, r)$ and $m = m(\omega, r)$ be the unique positive integers such that

$$\|d_x S^{-n}|E^s(x)\|^{-1} < r \le \|d_x S^{-(n-1)}|E^s(x)\|^{-1} \tag{8.7}$$

and

$$\|d_x T^m|E^u(x)\|^{-1} < r \le \|d_x T^{m-1}|E^u(x)\|^{-1}, \tag{8.8}$$

where $x = \pi(\omega)$. For each $q \in \mathbb{R}$, let J_q be the set of sequences $\omega \in \Sigma_A$ such that

$$-\lim_{r\to 0}\left(\frac{\sum_{k=0}^{n(\omega,r)-1} \psi^s(\sigma_-^k(\omega^-))}{\sum_{k=0}^{n(\omega,r)-1} d^s(\sigma_-^k(\omega^-))} + \frac{\sum_{k=0}^{m(\omega,r)-1} \psi^u(\sigma_+^k(\omega^+))}{\sum_{k=0}^{m(\omega,r)-1} d^u(\sigma_+^k(\omega^+))}\right) = \alpha(q),$$

where $\omega^- = \pi_-(\omega)$ and $\omega^+ = \pi_+(\omega)$.

Moreover, for each $q \in \mathbb{R}$, let μ_q^s and μ_q^u be respectively the equilibrium measures of the functions

$$-T_s(q)d^s + q\psi^s \quad \text{and} \quad -T_u(q)d^u + q\psi^u.$$

By (4.6), we obtain

$$0 = -T_s'(q)\int_{\Sigma_A^-} d^s\, d\mu_q^s + \int_{\Sigma_A^-} \psi^s\, d\mu_q^s$$

and

$$0 = -T_u'(q)\int_{\Sigma_A^+} d^u\, d\mu_q^u + \int_{\Sigma_A^+} \psi^u\, d\mu_q^u.$$

This implies that

$$\alpha_s(q) = -T_s'(q) = -\frac{\int_{\Sigma_A^-} \psi^s\, d\mu_q^s}{\int_{\Sigma_A^-} d^s\, d\mu_q^s}$$

and

$$\alpha_u(q) = -T_u'(q) = -\frac{\int_{\Sigma_A^+} \psi^u\, d\mu_q^u}{\int_{\Sigma_A^+} d^u\, d\mu_q^u}.$$

Since the measures μ_q^s and μ_q^u are ergodic, by Birkhoff's ergodic theorem we have

$$\lim_{n\to\infty} -\frac{\sum_{k=0}^{n-1} \psi^s(\sigma_-^k(\omega^-))}{\sum_{k=0}^{n-1} d^s(\sigma_-^k(\omega^-))} = \alpha_s(q)$$

for μ_q^s-almost every $\omega^- \in \Sigma_A^-$, and

$$\lim_{m\to\infty} -\frac{\sum_{k=0}^{m-1} \psi^u(\sigma_+^k(\omega^+))}{\sum_{k=0}^{m-1} d^u(\sigma_+^k(\omega^+))} = \alpha_u(q)$$

for μ_q^u-almost every $\omega^+ \in \Sigma_A^+$. Therefore, given $\omega \in \Sigma_A$ and $\delta > 0$, there exists an $r(\omega) > 0$ such that

$$\alpha_s(q) - \delta < -\frac{\sum_{k=0}^{n(\omega,r)-1} \psi^s(\sigma_-^k(\omega^-))}{\sum_{k=0}^{n(\omega,r)-1} d^s(\sigma_-^k(\omega^-))} < \alpha_s(q) + \delta \tag{8.9}$$

and

$$\alpha_u(q) - \delta < -\frac{\sum_{k=0}^{m(\omega,r)-1} \psi^u(\sigma_+^k(\omega^+))}{\sum_{k=0}^{m(\omega,r)-1} d^u(\sigma_+^k(\omega^+))} < \alpha_u(q) + \delta \tag{8.10}$$

for $r \in (0, r(\omega))$.

On the other hand, since μ_q^s and μ_q^u are equilibrium measures of Hölder continuous functions they are Gibbs measures. Moreover, it follows from (8.5) that

$$P_{\sigma_-|\Sigma_A^-}(-T_s(q)d^s + q\psi^s) = P_{\sigma_+|\Sigma_A^+}(-T_u(q)d^u + q\psi^u) = 0.$$

Hence, there exist constants $D_1, D_2 > 0$ such that

$$D_1 \leq \frac{\mu_q^s(C_{i_{-n}\cdots i_0}^-)}{\exp\left(-T_s(q)\sum_{k=0}^{n-1} d^s(\sigma_-^k(\omega^-)) + q\sum_{k=0}^{n-1} \psi^s(\sigma_-^k(\omega^-))\right)} \leq D_2 \tag{8.11}$$

and

$$D_1 \leq \frac{\mu_q^u(C_{i_0\cdots i_m}^+)}{\exp\left(-T_u(q)\sum_{k=0}^{m-1} d^u(\sigma_+^k(\omega^+)) + q\sum_{k=0}^{m-1} \psi^u(\sigma_+^k(\omega^+))\right)} \leq D_2, \tag{8.12}$$

for every $n, m \in \mathbb{N}$ and $\omega = (\cdots i_{-1} i_0 i_1 \cdots) \in \Sigma_A$, where

$$C_{i_{-n}\cdots i_0}^- \subset \Sigma_A^- \quad \text{and} \quad C_{i_0\cdots i_m}^+ \subset \Sigma_A^+$$

are cylinder sets.

Given $x \in Z$, let $R(x)$ be a rectangle of the Markov system that contains x. We have $R(x) = \pi(C_{i_0})$, where $x = \pi(\cdots i_0 \cdots)$. We also consider the measures

$$\nu_q^s = \pi_*(\mu_q^s | C_{i_0}^-) \quad \text{in} \quad A^s(x) = \pi(\pi_-^{-1} C_{i_0}^-) \cap R(x),$$

and

$$\nu_q^u = \pi_*(\mu_q^u | C_{i_0}^+) \quad \text{in} \quad A^u(x) = \pi(\pi_+^{-1} C_{i_0}^+) \cap R(x).$$

Finally, we define a product measure ν_q in $R(x) = [A^s(x), A^u(x)]$ by

$$\nu_q = \nu_q^s \times \nu_q^u,$$

using (3.3) to define the product structure. Given $l > 0$, consider the sets

$$Q_l = \{\omega \in J_q : r(\omega) \geq 1/l\}.$$

Clearly,

$$Q_l \subset Q_{l+1} \quad \text{and} \quad J_q = \bigcup_{l>0} Q_l. \tag{8.13}$$

Lemma 8.1 *For each $x \in Z$, we have*

$$\underline{d}_{\nu_q^s}(y) \geq T_s(q) + q(\alpha_s(q) - \delta)$$

for ν_q^s-almost every $y \in A^s(x) \cap \pi(J_q)$, and

$$\underline{d}_{\nu_q^u}(z) \geq T_u(q) + q(\alpha_u(q) - \delta)$$

for ν_q^u-almost every $z \in A^u(x) \cap \pi(J_q)$.

Proof of the lemma Given $\omega = (i_0 \cdots) \in \Sigma_A^+$ and $r \in (0,1)$, let

$$\Delta(\omega, r) = \pi\left(\pi_+^{-1} C^+_{i_0 \cdots i_m}\right),$$

where $n = m(\omega, r)$. We note that these sets intersect at most along their boundaries for each given r. Proceeding in a similar manner to that in (5.14) and (5.15), now along the unstable direction, and using (8.8), one can show that diam $\Delta(\omega, r) < r$, provided that the diameter of the Markov system is sufficiently small. Furthermore, since each set R_i is the closure of its interior, there exists a $\rho > 0$ such that R_i contains a ball B_i of radius ρ for $i = 1, \dots, k$. This implies that there exists a constant $\kappa \in (0,1)$ (independent of ω and r) such that each set $\Delta(\omega, r)$ contains a ball of radius κr (see [3] for details). Since the sets $\Delta(\omega, r)$ intersect at most along their boundaries, it follows from elementary geometry that there exists a constant $C > 0$ (independent of r) such that each ball $B(x, r)$ intersects at most C of the sets $\Delta(\omega, r)$.

Now we construct a special cover of $\pi(Q_l)$. For each $\omega \in Q_l$, let $\tilde{\Delta}(\omega, r)$ be the largest set containing $\pi(\omega)$ such that:

1. $\tilde{\Delta}(\omega, r) = \Delta(\omega', r)$ for some $\omega' \in Q_l$ with $\pi(\omega') \in \tilde{\Delta}(\omega, r)$;
2. $\Delta(\omega', r) \subset \tilde{\Delta}(\omega, r)$ whenever $\pi(\omega') \in \tilde{\Delta}(\omega, r)$.

By construction, for a given r the sets $\tilde{\Delta}(\omega, r)$ form a cover of $\pi(Q_l)$. Let $\tilde{\Delta}_j = \Delta(\omega_j, r)$ with $\omega_j \in Q_l$, for $j = 1, \dots, N(r)$, be the elements of this cover. Given $r < 1/l$, it follows from (8.8), (8.10) and (8.12) that

$$\begin{aligned}
&\nu_q^u(B(x,r) \cap \pi(Q_l)) \\
&\quad \le \sum_{\tilde{\Delta}_j \cap B(x,r) \neq \varnothing} \nu_q^u(\Delta(\omega_j, r)) \\
&\quad \le D_2 \sum_{\tilde{\Delta}_j \cap B(x,r) \neq \varnothing} \|d_{\pi(\omega_j)} T^{m(\omega_j, r)} | E^u(\pi(\omega_j))\|^{-T_u(q)} \\
&\quad\quad \times \exp\left(q \sum_{k=0}^{m(\omega_j, r)-1} \psi^u(\sigma_+^k(\omega^+))\right) \\
&\quad \le D_2 \sum_{\tilde{\Delta}_j \cap B(x,r) \neq \varnothing} \|d_{\pi(\omega_j)} T^{m(\omega_j, r)} | E^u(\pi(\omega_j))\|^{-T_u(q) - q(\alpha_u(q) - \delta)}.
\end{aligned}$$

Using (8.8) again, we conclude that there exists a $C' > 0$ such that

$$\nu_q^u(B(x,r) \cap \pi(Q_l)) \le C' r^{T_u(q) + q(\alpha_u(q) - \delta)} \tag{8.14}$$

for every $x \in J$ and $r \in (0, 1/l)$. By the Borel density lemma, for ν_q^u-almost every $x \in \pi(Q_l)$ we have

$$\lim_{r \to 0} \frac{\nu_q^u(B(x,r) \cap \pi(Q_l))}{\nu_q^u(B(x,r))} = 1,$$

and thus, there exists a $\rho(x) > 0$ such that

$$\nu_q^u(B(x,r)) \le 2\nu_q^u(B(x,r) \cap \pi(Q_l))$$

for every $r \in (0, \rho(x))$. Together with (8.14) this implies that

$$\begin{aligned}\underline{d}_{\nu_q^u}(x) &= \liminf_{r\to 0} \frac{\log \nu_q^u(B(x,r))}{\log r} \\ &\ge \liminf_{r\to 0} \frac{\log \nu_q^u(B(x,r)\cap \pi(Q_l))}{\log r} \\ &\ge T_u(q) + q(\alpha_u(q) - \delta)\end{aligned}$$

for ν_q^u-almost every $x \in \pi(Q_l)$. By (8.13), we conclude that

$$\underline{d}_{\nu_q^u}(x) \ge T_u(q) + q(\alpha_u(q) - \delta)$$

for ν_q^u-almost every $x \in \pi(J_q)$. Since δ is arbitrary, this yields the desired result along the unstable direction. The corresponding result along the stable direction can be obtained in a similar manner. □

It follows from (8.13) and the arbitrariness of δ that

$$\underline{d}_{\nu_q^s}(y) \ge T_s(q) + q\alpha_s(q)$$

for ν_q^s-almost every $y \in A^s(x) \cap \pi(J_q)$, and

$$\underline{d}_{\nu_q^u}(z) \ge T_u(q) + q\alpha_u(q)$$

for ν_q^u-almost every $z \in A^u(x) \cap \pi(J_q)$. Since $\nu_q = \nu_q^s \times \nu_q^u$, we obtain

$$\begin{aligned}\underline{d}_{\nu_q}(x) &= \liminf_{r\to 0} \frac{\log \nu_q(B(x,r))}{\log r} \\ &\ge \underline{d}_{\nu_q^s}(x) + \underline{d}_{\nu_q^u}(x) \\ &\ge T_s(q) + T_u(q) + q(\alpha_s(q) + \alpha_u(q))\end{aligned}$$

for ν_q-almost every $x \in \pi(J_q)$. It follows from Proposition 4.7 that

$$\dim_H \pi(J_q) \ge T(q) + q\alpha(q). \tag{8.15}$$

Lemma 8.2 *For each $x \in \pi(J_q)$, we have*

$$\overline{d}_{\nu_q}(x) \le T(q) + q\alpha(q).$$

Proof of the lemma By the choice of n and m in (8.7) and (8.8), there exists a constant $c > 0$ such that

$$D(\omega, r) = \bigcup_{j=-n(\omega,r)}^{m(\omega,r)} S^{-k} R_{i_j} \subset B(x, cr) \tag{8.16}$$

for every $x = \pi(\omega) \in \Lambda$ and $r \in (0,1)$. It follows from (8.11) and (8.12) that for every $x = \pi(\omega)$ with $\omega \in Q_l$ and $r < 1/l$, taking $n = n(\omega, r)$ and $m = m(\omega, r)$ we obtain

$$\begin{aligned}
\nu_q(B(x, cr)) &\geq \nu_q(D(\omega, r)) = \mu_q^s(C^-_{i_0\cdots i_n})\mu_q^u(C^+_{i_0\cdots i_m}) \\
&\geq D_1^2 \exp\left(-T_s(q)\sum_{k=0}^{n-1} d^s(\sigma_-^k(\omega^-)) + q\sum_{k=0}^{n-1} \psi^s(\sigma_-^k(\omega^-))\right) \\
&\quad \times \exp\left(-T_u(q)\sum_{k=0}^{m-1} d^u(\sigma_+^k(\omega^+)) + q\sum_{k=0}^{m-1} \psi^u(\sigma_+^k(\omega^+))\right) \\
&= D_1^2 \exp\left(-T_s(q)\log\|d_x S^{-n}|E^s(x)\| + q\sum_{k=0}^{n-1} I_g(S^{-k}(x))\right) \\
&\quad \times \exp\left(-T_u(q)\log\|d_x T^{m}|E^u(x)\| + q\sum_{k=0}^{m-1} I_g(T^{k}(x))\right).
\end{aligned}$$

Therefore, by (8.7), (8.8), (8.9) and (8.10), we have

$$\begin{aligned}
\overline{d}_{\nu_q}(x) &= \limsup_{r\to 0} \frac{\log \nu_q(B(x,r))}{\log r} \\
&\leq T_s(q) \limsup_{r\to 0} \frac{-\log\|d_x S^{-n(\omega,r)}|E^s(x)\|}{\log r} \\
&\quad + T_u(q) \limsup_{r\to 0} \frac{-\log\|d_x T^{m(\omega,r)}|E^u(x)\|}{\log r} \\
&\quad + \limsup_{r\to 0} \frac{q\sum_{k=-(n(\omega,r)-1)}^{m(\omega,r)-1} I_g(S^k(x))}{\log r} \\
&\leq T_s(q) + T_u(q) + q(\alpha_s(q) + \alpha_u(q) + 2\delta)
\end{aligned}$$

for every $x \in \pi(Q_l)$. It follows from (8.13) and the arbitrariness of δ that

$$\overline{d}_{\nu_q}(x) \leq T_s(q) + T_u(q) + q(\alpha_s(q) + \alpha_u(q))$$

for every $x \in \pi(J_q)$. □

By Lemma 8.2, it follows from Proposition 4.7 that

$$\dim_H \pi(J_q) \leq T_s(q) + T_u(q) + q(\alpha_s(q) + \alpha_u(q)).$$

Together with (8.15) this implies that

$$\dim_H \pi(J_q) = T_s(q) + T_u(q) + q(\alpha_s(q) + \alpha_u(q)). \tag{8.17}$$

Lemma 8.3 *Given $\gamma > 0$, there exists a $K > 0$ such that*

$$\nu(B(y, \gamma r)) \leq K\nu(B(y, r))$$

for every $y \in R(x)$ and any sufficiently small $r > 0$.

Proof of the lemma For ν-almost every $y \in \Lambda$, let ν_y^s and ν_y^u be respectively the conditional measures of ν in $A^s(x)$ and $A^u(x)$. Repeating verbatim the arguments in the case of repellers of conformal maps (see Lemma 6.1.5 in [3]), we find that there exists a $C > 0$ such that

$$\nu_y^s(B^s(y, 2r)) \leq C\nu_y^s(B^s(y, r)) \quad \text{and} \quad \nu_y^u(B^u(y, 2r)) \leq C\nu_y^u(B^u(y, r)) \tag{8.18}$$

for every $y \in \Lambda$ and any sufficiently small $r > 0$, where $B^s(y, r)$ and $B^u(y, r)$ are the open balls centered at y of radius r with respect to the distances induced respectively on the local stable and unstable manifolds $V^s(y)$ and $V^u(y)$.

Now we observe that there exists a $\kappa > 1$ such that

$$\Lambda \cap B(y, \gamma r) \subset \left[\Lambda \cap B^s(y, \kappa r), \Lambda \cap B^u(y, \kappa r)\right] \tag{8.19}$$

and

$$\left[\Lambda \cap B^s(y, r/\kappa), \Lambda \cap B^u(y, r/\kappa)\right] \subset \Lambda \cap B(y, r) \tag{8.20}$$

for every $y \in \Lambda$ and any sufficiently small $r > 0$. It follows from (8.19) that

$$\nu(B(y, \gamma r)) \leq c\nu_y^u(B^u(y, \kappa r))\nu_y^s(B^s(y, \kappa r))$$

for some constant $c > 0$ (independent of y and r). Applying (8.18) a number n of times such that $\kappa 2^{-n} < 1/\kappa$, we obtain

$$\nu(B(y, \gamma r)) \leq C^{2n}\nu_y^u(B^u(y, r/\kappa))\nu_y^s(B^s(y, r/\kappa)).$$

Hence, it follows from (8.20) that

$$\nu(B(y, \gamma r)) \leq cC^{2n}\nu(B(y, r))$$

for some constant $c > 0$ (independent of y and r) and the desired inequality holds with $K = cC^{2n}$. □

The following result relates the level sets K_α to the sets J_q.

Lemma 8.4 *We have* $\pi(J_q) = K_{\alpha(q)} \cap Z$.

Proof of the lemma Let $r \in (0,1)$ and take $n = n(\omega, r)$ and $m = m(\omega, r)$ as in (8.7) and (8.8). Proceeding as in the proof of Lemma 8.1, we find that there exists a $\kappa > 0$ (independent of r) such that for each $x = \pi(\omega) \in Z$ there exists a $y \in D(\omega, r)$ (see (8.16)) for which

$$B(y, \kappa r) \subset D(\omega, r) \subset B(x, cr). \tag{8.21}$$

Moreover, $B(x,r) \subset B(y, dr)$ for some constant $d > 0$ (independent of x and r). By Lemma 8.3, we obtain

$$\begin{aligned} \nu(D(\omega,r)) &\le \nu(B(x,cr)) \le K_1 \nu(B(x,r)) \\ &\le K_1 \nu(B(y,dr)) \le K_2 \nu(B(y,\kappa r)) \\ &\le K_2 \nu(D(\omega,r)) \end{aligned}$$

for some constants $K_1, K_2 > 0$. This implies that if either of the two limits

$$\lim_{r\to 0} \frac{\log \nu(D(\omega,r))}{\log r} \quad \text{and} \quad \lim_{r\to 0} \frac{\log \nu(B(x,r))}{\log r} \tag{8.22}$$

exists, then the other also exists and has the same value. On the other hand, since ν is the equilibrium measure of g and $P_{\Phi|\Lambda}(g) = 0$, if the first limit exists, then

$$a := \lim_{r\to 0} \frac{\log \nu(D(\omega,r))}{\log r} = \lim_{r\to 0} \frac{\sum_{k=-(n(\omega,r)-1)}^{m(\omega,r)-1} I_g(S^k(x))}{\log r}.$$

It follows from (8.7) and (8.8) that $x = \pi(\omega) \in \pi(J_q)$ if and only if

$$\begin{aligned} a &= \lim_{r\to 0} \left(\frac{\sum_{k=0}^{n(\omega,r)-1} I_g(f^{-k}(x))}{\log r} + \frac{\sum_{k=0}^{m(\omega,r)-1} I_g(f^k(x))}{\log r} \right) \\ &= \lim_{r\to 0} \left(\frac{\sum_{k=0}^{n(\omega,r)-1} I_g(S^{-k}(x))}{-\log \|d_x S^{-n(\omega,r)}|E^s(x)\|} + \frac{\sum_{k=0}^{m(\omega,r)-1} I_g(T^k(x))}{-\log \|d_x T^{m(\omega,r)}|E^u(x)\|} \right) = \alpha(q). \end{aligned}$$

Hence, $x \in \pi(J_q)$ if and only if the second limit in (8.22) is equal to $\alpha(q)$. □

By (8.17) and Lemma 8.4, in view of (8.6) we obtain

$$\begin{aligned} \dim_H K_{\alpha(q)} &= 1 + \dim_H(K_{\alpha(q)} \cap Z) \\ &= 1 + T_s(q) + T_u(q) + q(\alpha_s(q) + \alpha_u(q)) \\ &= T(q) + q\alpha(q). \end{aligned}$$

This completes the proof of the theorem. □

It is also shown in [82] that if ν is not a measure of full dimension, that is, $\dim_H \nu \ne \dim_H \Lambda$, then the functions $\mathcal{D}$ and T are analytic and strictly convex.

8.3 Entropy Spectra and Cohomology

In this section we consider the particular case of entropy spectra for hyperbolic flows. We emphasize that unlike in Chap. 7, the results for these spectra cannot be obtained from the results for dimension spectra in Sect. 8.2 (in Chap. 7 we were instead considering BS-dimension spectra, not dimension spectra). In particular, we describe appropriate versions of the results in Sect. 7.4. We also show that the cohomology assumptions required in the study of the irregular sets are generically satisfied.

Let Φ be a C^1 flow with a locally maximal hyperbolic set Λ and let $g\colon \Lambda \to \mathbb{R}$ be a continuous function. For each $\alpha \in \mathbb{R}$, we consider the set

$$K_\alpha = \left\{ x \in \Lambda : \lim_{t\to\infty} \frac{1}{t} \int_0^t g(\varphi_\tau(x))\, d\tau = \alpha \right\}.$$

One can easily verify that K_α is Φ-invariant.

We recall that a function $g\colon \Lambda \to \mathbb{R}$ is said to be Φ-cohomologous to a function $h\colon \Lambda \to \mathbb{R}$ if there exists a bounded measurable function $q\colon \Lambda \to \mathbb{R}$ such that

$$g(x) - h(x) = \lim_{t\to 0} \frac{q(\varphi_t(x)) - q(x)}{t}$$

for every $x \in \Lambda$. In particular, if $g\colon \Lambda \to \mathbb{R}$ is Φ-cohomologous to a constant $c \in \mathbb{R}$ in Λ, then

$$\begin{aligned}\left| \frac{1}{t} \int_0^t g(\varphi_\tau(x))\, d\tau - c \right| &= \frac{1}{t} \lim_{s\to 0} \frac{1}{s} \left| \int_s^{s+t} q(\varphi_\tau(x))\, d\tau - \int_0^t q(\varphi_\tau(x))\, d\tau \right| \\ &= \frac{1}{t} \lim_{s\to 0} \frac{1}{s} \left| \int_t^{s+t} q(\varphi_\tau(x))\, d\tau - \int_0^s q(\varphi_\tau(x))\, d\tau \right| \\ &\le \frac{2\sup|q|}{t} \end{aligned} \tag{8.23}$$

for every $x \in \Lambda$ and $t > 0$, and hence, $K_c = \Lambda$. This shows that it is only interesting to consider the case when g is not cohomologous to a constant.

Now we introduce the entropy spectrum. Given $\alpha \in \mathbb{R}$, let

$$\mathcal{E}(\alpha) = h(\Phi|K_\alpha).$$

The function $\mathcal{E}$ is called the *entropy spectrum for the Birkhoff averages of* g. For each $q \in \mathbb{R}$, let ν_q be the equilibrium measure for qg and write

$$T(q) = P_\Phi(qg),$$

where $P_\Phi(qg)$ is the topological pressure of qg with respect to Φ.

The following is a multifractal analysis of the spectrum $\mathcal{E}$.

Theorem 8.4 ([12]) *Let Φ be a $C^{1+\delta}$ flow with a locally maximal hyperbolic set Λ such that $\Phi|\Lambda$ is topologically mixing and let $g\colon \Lambda \to \mathbb{R}$ be a Hölder continuous function with $P_\Phi(g) = 0$. Then the following properties hold*:

1. *the domain of $\mathcal{E}$ is a closed interval in $[0, \infty)$ coinciding with the range of the function $\alpha = -T'$, and for each $q \in \mathbb{R}$ we have*

$$\mathcal{E}(\alpha(q)) = T(q) + q\alpha(q) = h_{\nu_q}(\Phi|\Lambda);$$

2. *if g is not Φ-cohomologous to a constant in Λ, then the functions $\mathcal{E}$ and T are analytic and strictly convex.*

Proof Consider a Markov system for Φ on Λ, the associated suspension flow Ψ and the coding map $\pi : Y \to \Lambda$ defined by (3.11). It follows from (3.12) that

$$\lim_{t\to\infty} \frac{1}{t} \int_0^t (g \circ \pi)(\psi_\tau(x))\,d\tau = \alpha$$

if and only if

$$\lim_{t\to\infty} \frac{1}{t} \int_0^t g(\varphi_\tau(\pi(x)))\,d\tau = \alpha.$$

This shows that $\mathcal{E} = \mathcal{D}_1$, with $\mathcal{D}_u$ as in Sect. 7.1. Hence, the desired result follows from Theorem 7.1 taking $u = 1$. □

Given a continuous function $g\colon \Lambda \to \mathbb{R}$, the *irregular set for the Birkhoff averages of* g (with respect to Φ) is defined by

$$\mathcal{B}(g) = \left\{ x \in \Lambda : \liminf_{t\to\infty} \frac{1}{t} \int_0^t g(\varphi_\tau(x))\,d\tau < \limsup_{t\to\infty} \frac{1}{t} \int_0^t g(\varphi_\tau(x))\,d\tau \right\}.$$

One can easily verify that $\mathcal{B}(g)$ is Φ-invariant. By Birkhoff's ergodic theorem, the set $\mathcal{B}(g)$ has zero measure with respect to any Φ-invariant finite measure. Moreover, by (8.23), if g is Φ-cohomologous to a constant in Λ, then $\mathcal{B}(g) = \varnothing$.

The following result gives a necessary and sufficient condition so that the irregular set has full topological entropy.

Theorem 8.5 ([12]) *Let Φ be a $C^{1+\delta}$ flow with a locally maximal hyperbolic set Λ such that $\Phi|\Lambda$ is topologically mixing and let $g\colon \Lambda \to \mathbb{R}$ be a Hölder continuous function. Then the following properties are equivalent*:

1. *g is not Φ-cohomologous to a constant in Λ;*
2. *$h(\Phi|\mathcal{B}(g)) = h(\Phi|\Lambda)$.*

Proof If g is Φ-cohomologous to a constant, then $\mathcal{B}(g) = \varnothing$.

Now we assume that g is not Φ-cohomologous to a constant. Consider a Markov system, the associated suspension flow Ψ and the coding map $\pi : Y \to \Lambda$. It follows from (3.12) that

$$\liminf_{t\to\infty}\frac{1}{t}\int_0^t (g\circ\pi)(\psi_\tau(x))\,d\tau < \limsup_{t\to\infty}\frac{1}{t}\int_0^t (g\circ\pi)(\psi_\tau(x))\,d\tau$$

if and only if

$$\liminf_{t\to\infty}\frac{1}{t}\int_0^t g(\varphi_\tau(\pi(x)))\,d\tau < \limsup_{t\to\infty}\frac{1}{t}\int_0^t g(\varphi_\tau(\pi(x)))\,d\tau.$$

Therefore,

$$\mathcal{B}(g)=\pi(\mathcal{B}(g\circ\pi)). \tag{8.24}$$

To complete the proof we proceed as in [17]. Let $R\subset\Lambda$ be the set of points $y\in\Lambda$ such that $\varphi_t(x)$ is on the boundary of some element of the Markov system for some $t\in\mathbb{R}$. We note that R is Φ-invariant and that

$$\pi:\pi^{-1}(\Lambda\setminus R)\to\Lambda\setminus R$$

is a homeomorphism. Moreover, since there exist cylinder sets $C\subset X$ such that $\pi(C)$ is disjoint from R, we have

$$h(\Psi|\pi^{-1}R)<h(\Psi)\quad\text{and}\quad h(\Phi|R)<h(\Phi|\Lambda).$$

By (8.24), we obtain

$$h(\Phi|\mathcal{B}(g))=h(\Psi|\mathcal{B}(g\circ\pi)),$$

and it follows from Theorem 7.4 that

$$h(\Phi|\Lambda)=h(\Psi)=h(\Psi|\mathcal{B}(g\circ\pi))=h(\Phi|\mathcal{B}(g)).$$

This completes the proof of the theorem. □

Theorem 8.5 is a counterpart of results of Barreira and Schmeling in [17] for discrete time.

Now we show that most Hölder continuous functions are not Φ-cohomologous to a constant. Let $C^\gamma(\Lambda)$ be the space of Hölder continuous functions in Λ with Hölder exponent $\gamma\in(0,1)$. We define the norm of a function $\varphi\in C^\gamma(\Lambda)$ by

$$\|\varphi\|_\gamma=\sup\{|\varphi(x)|:x\in\Lambda\}+\sup\left\{\frac{|\varphi(x)-\varphi(y)|}{d(x,y)^\gamma}:x,y\in\Lambda\text{ and }x\neq y\right\}, \tag{8.25}$$

where d is the distance on M. We recall that Φ is said to be *topologically transitive* on Λ (or simply $\Phi|\Lambda$ is said to be *topologically transitive*) if for any nonempty open sets U and V intersecting Λ there exist a $t\in\mathbb{R}$ such that $\varphi_t(U)\cap V\cap\Lambda\neq\varnothing$.

Theorem 8.6 ([12]) *Let Φ be a C^1 flow with a locally maximal hyperbolic set Λ such that $\Phi|\Lambda$ is topologically transitive. For each $\gamma\in(0,1)$, the set of all functions in $C^\gamma(\Lambda)$ that are not Φ-cohomologous to a constant is open and dense in $C^\gamma(\Lambda)$.*

Proof Let

$$G=\{g\in C^\gamma(\Lambda): g \text{ is not } \Phi\text{-cohomologous to a constant}\}$$

and take $g\in G$. By Livschitz's theorem (see Theorem 19.2.4 in [62]), there exist points $x_i=\varphi_{T_i}(x_i)$ for $i=0,1$ for which

$$c=\left|\frac{1}{T_0}\int_0^{T_0} g(\varphi_\tau(x_0))\,d\tau-\frac{1}{T_1}\int_0^{T_1} g(\varphi_\tau(x_1))\,d\tau\right|\neq 0.$$

For each $f\in C^\gamma(\Lambda)$ such that $\|f-g\|_\gamma<c/2$, we have

$$\left|\frac{1}{T_i}\int_0^{T_i}(f-g)(\varphi_\tau(x_i))\,d\tau\right|\le\sup\{|f(x)-g(x)|: x\in\Lambda\}\le\|f-g\|_\gamma<\frac{c}{2}$$

for $i=0,1$, and hence,

$$\frac{1}{T_0}\int_0^{T_0} f(\varphi_\tau(x_0))\,d\tau\neq\frac{1}{T_1}\int_0^{T_1} f(\varphi_\tau(x_1))\,d\tau.$$

This implies that f is not Φ-cohomologous to a constant, and thus, the set G is open.

Now let Γ_0 and Γ_1 be distinct periodic orbits, and let $h\in C^\gamma(\Lambda)$ be a Hölder continuous function such that $h|\Gamma_i=i$ for $i=0,1$. Take $g\notin G$. For each $\varepsilon>0$, the function $g_\varepsilon=g+\varepsilon h\in C^\gamma(\Lambda)$ is not Φ-cohomologous to a constant, because the averages on Γ_0 and Γ_1 differ by ε. Moreover,

$$\|g_\varepsilon-g\|_\gamma\le\varepsilon\|h\|_\gamma,$$

and hence the function g can be arbitrarily approximated by functions in G. Therefore, G is dense in $C^\gamma(\Lambda)$. □

Theorems 8.5 and 8.6 readily imply the following result.

Theorem 8.7 *Let Φ be a $C^{1+\delta}$ flow with a locally maximal hyperbolic set Λ such that $\Phi|\Lambda$ is topologically mixing. Given $\gamma\in(0,1)$, for an open and dense family of functions $g\in C^\gamma(\Lambda)$ we have $h(\Phi|\mathcal{B}(g))=h(\Phi|\Lambda)$.*

Part IV
Variational Principles

This final part is dedicated to the study of conditional variational principles. This corresponds to describing the topological entropy or the Hausdorff dimension of the level sets of a given function. In Chap. 9 we obtain a conditional variational principle for flows with a locally maximal hyperbolic set and we study the analyticity of several classes of multifractal spectra. In particular, we consider spectra defined by local entropies and Lyapunov exponents. In Chap. 10 we obtain a multidimensional version of multifractal analysis for hyperbolic flows. This corresponds to computing the topological entropy of the multidimensional level sets associated to several Birkhoff averages. In Chap. 11 we establish a conditional variational principle for the dimension spectra of Birkhoff averages, considering simultaneously averages into the future and into the past.

Chapter 9
Entropy Spectra

In this chapter we establish a conditional variational principle for flows with a locally maximal hyperbolic set. In other words, we express the topological entropy of the level sets of the Birkhoff averages of a given function in terms of a conditional variational principle. As an application of this principle, we establish the analyticity of several classes of multifractal spectra for hyperbolic flows. In particular, we consider the multifractal spectra for the local entropies and for the Lyapunov exponents.

9.1 A Conditional Variational Principle

This section is dedicated to establishing a conditional variational principle for hyperbolic flows. We consider multifractal spectra obtained from ratios of Birkhoff averages.

Let $\Phi = \{\varphi_t\}_{t\in\mathbb{R}}$ be a continuous flow and let Λ be a Φ-invariant set. We denote by $C(\Lambda)$ the space of all continuous functions $a\colon \Lambda \to \mathbb{R}$. Given $a, b \in C(\Lambda)$ with $b > 0$ and $\alpha \in \mathbb{R}$, let

$$K_\alpha = K_\alpha(a,b) = \left\{ x \in \Lambda : \lim_{t\to\infty} \frac{\int_0^t a(\varphi_s(x))\,ds}{\int_0^t b(\varphi_s(x))\,ds} = \alpha \right\}.$$

One can easily verify that the set K_α is Φ-invariant.

Now we introduce the entropy spectrum.

Definition 9.1 The function $\mathcal{F} = \mathcal{F}^{(a,b)}$ defined by

$$\mathcal{F}(\alpha) = h(\Phi|K_\alpha) \tag{9.1}$$

is called the *entropy spectrum* for the pair of functions (a, b).

L. Barreira, *Dimension Theory of Hyperbolic Flows*,
Springer Monographs in Mathematics, DOI 10.1007/978-3-319-00548-5_9,

The following result is a conditional variational principle for the spectrum $\mathcal{F}$. Let

$$\underline{\alpha} = \inf\left\{\frac{\int_\Lambda a\,d\mu}{\int_\Lambda b\,d\mu} : \mu \in \mathcal{M}\right\}$$

and

$$\overline{\alpha} = \sup\left\{\frac{\int_\Lambda a\,d\mu}{\int_\Lambda b\,d\mu} : \mu \in \mathcal{M}\right\},$$

where $\mathcal{M}$ is the set of all Φ-invariant probability measures on Λ.

Theorem 9.1 ([15]) *Let Φ be a C^1 flow with a locally maximal hyperbolic set Λ such that $\Phi|\Lambda$ is topologically mixing and let $a, b \colon \Lambda \to \mathbb{R}$ be Hölder continuous functions with $b > 0$. Then the following properties hold:*

1. *if $\alpha \notin [\underline{\alpha}, \overline{\alpha}]$, then $K_\alpha = \varnothing$;*
2. *if $\alpha \in (\underline{\alpha}, \overline{\alpha})$, then $K_\alpha \neq \varnothing$,*

$$\mathcal{F}(\alpha) = \max\left\{h_\mu(\Phi) : \frac{\int_\Lambda a\,d\mu}{\int_\Lambda b\,d\mu} = \alpha \text{ and } \mu \in \mathcal{M}\right\} \tag{9.2}$$

and

$$\mathcal{F}(\alpha) = \min\{P_\Phi(qa - q\alpha b) : q \in \mathbb{R}\}. \tag{9.3}$$

Proof Let us assume that $K_\alpha \neq \varnothing$ and take $x \in K_\alpha$. The sequence of probability measures $(\mu_n)_{n\in\mathbb{N}}$ in Λ such that

$$\int_\Lambda u\,d\mu_n = \frac{1}{n}\int_0^n u(\varphi_s(x))\,ds$$

for every $u \in C(\Lambda)$ has an accumulation point $\mu \in \mathcal{M}$. Therefore,

$$\begin{aligned}\alpha &= \lim_{t\to\infty} \frac{\int_0^t a(\varphi_s(x))\,ds}{\int_0^t b(\varphi_s(x))\,ds}\\ &= \lim_{n\to\infty} \frac{\int_\Lambda a\,d\mu_n}{\int_\Lambda b\,d\mu_n}\\ &= \frac{\int_\Lambda a\,d\mu}{\int_\Lambda b\,d\mu} \in [\underline{\alpha}, \overline{\alpha}].\end{aligned}$$

This establishes the first property.

Now we establish the second property. By Proposition 4.2, for each $\alpha \in \mathbb{R}$ we have

$$\inf_{q\in\mathbb{R}} P_\Phi(qa - q\alpha b) = \inf_{q\in\mathbb{R}} \sup\left\{h_\mu(\Phi) + \int_\Lambda (qa - q\alpha b)\,d\mu : \mu \in \mathcal{M}\right\}$$

$$\geq \sup\left\{h_\mu(\Phi) : \frac{\int_\Lambda a\,d\mu}{\int_\Lambda b\,d\mu} = \alpha \text{ and } \mu \in \mathcal{M}\right\}. \tag{9.4}$$

On the other hand, given $\alpha \in (\underline{\alpha}, \overline{\alpha})$, there exist measures $\nu_-, \nu_+ \in \mathcal{M}$ such that

$$\frac{\int_\Lambda a\,d\nu_-}{\int_\Lambda b\,d\nu_-} < \alpha < \frac{\int_\Lambda a\,d\nu_+}{\int_\Lambda b\,d\nu_+}.$$

Moreover, for each $q \in \mathbb{R}$ and $\mu \in \mathcal{M}$, we have

$$P_\Phi(qa - q\alpha b) \geq h_\mu(\Phi) + q\left(\int_\Lambda a\,d\mu - \alpha\int_\Lambda b\,d\mu\right),$$

and hence,

$$\liminf_{q\to\pm\infty} P_\Phi(qa - q\alpha b) \geq h_{\nu_\pm}(\Phi) + \liminf_{q\to\pm\infty} q\left(\int_\Lambda a\,d\nu_\pm - \alpha\int_\Lambda b\,d\nu_\pm\right) = +\infty.$$

In particular, the map $q \mapsto P_\Phi(qa - q\alpha b)$ attains its infimum at some point $q = q(\alpha) \in \mathbb{R}$. We note that this map is analytic (see [92]). Denoting by μ_α the equilibrium measure for the function $q(\alpha)(a - \alpha b)$, we obtain

$$0 = \frac{d}{dq} P_\Phi(qa - q\alpha b)\Big|_{q=q(\alpha)} = \int_\Lambda a\,d\mu_\alpha - \alpha\int_\Lambda b\,d\mu_\alpha.$$

Therefore, inequality (9.4) is in fact an equality and

$$\begin{aligned}\inf_{q\in\mathbb{R}} P_\Phi(qa - q\alpha b) &= h_{\mu_\alpha}(\Phi)\\ &= \max\left\{h_\mu(\Phi) : \frac{\int_\Lambda a\,d\mu}{\int_\Lambda b\,d\mu} = \alpha \text{ and } \mu \in \mathcal{M}\right\}.\end{aligned} \tag{9.5}$$

Since Λ is hyperbolic, the flow $\Phi|\Lambda$ is expansive, in which case identity (4.4) takes the simpler form

$$h_\mu(\Phi) = \inf\{h(\Phi|Z) : \mu(Z) = 1\}$$

(that is, the limits in ε in (4.4) are not necessary provided that ε is sufficiently small). Since μ_α is ergodic, we have $\mu_\alpha(K_\alpha) = 1$ and thus,

$$h_{\mu_\alpha}(\Phi) \leq \mathcal{F}(\alpha). \tag{9.6}$$

In view of (9.5) and (9.6), it remains to show that

$$\mathcal{F}(\alpha) \leq \inf_{q\in\mathbb{R}} P_\Phi(qa - q\alpha b).$$

Otherwise, there would exist $q \in \mathbb{R}$, $\delta > 0$ and $c > 0$ such that

$$\mathcal{F}(\alpha) - \delta > c > P_\Phi(qa - q\alpha b). \tag{9.7}$$

Let $u = qa - \alpha qb$ and

$$K_{\alpha,\delta,\tau} = \left\{ x \in \Lambda : \left| \int_0^t u(\varphi_s(x))\, ds \right| < \delta t \text{ for } t \geq \tau \right\}.$$

We have

$$K_\alpha \subset \bigcup_{\tau \in \mathbb{N}} K_{\alpha,\delta,\tau} = K_{\alpha,\delta}$$

and it follows from the basic properties of the topological entropy (see [81]) that

$$\begin{aligned} \lim_{\tau \to +\infty} h(\Phi | K_{\alpha,\delta,\tau}) &= h(\Phi | K_{\alpha,\delta}) \\ &\geq h(\Phi | K_\alpha) = \mathcal{F}(\alpha). \end{aligned}$$

In particular, there exists a $\tau \in \mathbb{N}$ such that

$$c + \delta < h(\Phi | K_{\alpha,\delta,\tau}). \tag{9.8}$$

For each $y \in B(x,t,\varepsilon)$ and $s \in [0,t]$, we have $d(\varphi_s(x), \varphi_s(y)) < \varepsilon$ and thus,

$$|u(x,t,\varepsilon)| \leq \left| \int_0^t u(\varphi_s(y))\, ds \right| + \eta(\varepsilon) t,$$

where

$$\eta(\varepsilon) = \sup\bigl\{ |u(x) - u(y)| : d(x,y) < \varepsilon \bigr\}.$$

Moreover, if $B(x,t,\varepsilon) \cap K_{\alpha,\delta,\tau} \neq \varnothing$, then there exists a $y \in B(x,t,\varepsilon)$ such that

$$\left| \int_0^t u(\varphi_s(y))\, ds \right| < \delta t$$

whenever $t \geq \tau$. This implies that

$$|u(x,t,\varepsilon)| \leq [\delta + \eta(\varepsilon)] t$$

whenever $B(x,t,\varepsilon) \cap K_{\alpha,\delta,\tau} \neq \varnothing$ and $t \geq \tau$. Hence,

$$\begin{aligned} M(K_{\alpha,\delta,\tau}, u, c, \varepsilon) &= \lim_{T \to \infty} \inf_{\Gamma} \sum_{(x,t) \in \Gamma} \exp(u(x,t,\varepsilon) - ct) \\ &\geq \lim_{T \to \infty} \inf_{\Gamma} \sum_{(x,t) \in \Gamma} \exp(-[\delta + \eta(\varepsilon)] t - ct) \\ &= M\bigl(K_{\alpha,\delta,\tau}, 0, c + \delta + \eta(\varepsilon), \varepsilon\bigr), \end{aligned}$$

where the infimum is taken over all finite or countable sets $\Gamma = \{(x_i, t_i)\}_{i \in I}$ such that $(x_i, t_i) \in X \times [T, \infty)$ for $i \in I$, and $\bigcup_{i \in I} B(x_i, t_i, \varepsilon) \supset K_{\alpha,\delta,\tau}$. Since u is continuous, we have $\eta(\varepsilon) \to 0$ when $\varepsilon \to 0$, and in view of (9.8) it follows from the definition of the topological entropy that

$$M(K_{\alpha,\delta,\tau}, u, c, \varepsilon) > 0$$

for any sufficiently small $\varepsilon > 0$. Therefore,

$$c \le P_{\Phi|K_{\alpha,\delta,\tau}}(qa - q\alpha b) \le P_\Phi(qa - q\alpha b),$$

which contradicts (9.7). This completes the proof of the theorem. □

Identity (9.2) is called a *conditional variational principle* for the entropy spectrum. Theorem 9.2 below gives a necessary and sufficient condition in terms of the functions a and b so that $\underline{\alpha} < \overline{\alpha}$.

We also explain how to obtain a measure at which the maximum in (9.2) is attained. Let $q(\alpha) \in \mathbb{R}$ be a point where the function $q \mapsto P_\Phi(qa - q\alpha b)$ attains its infimum (it is shown in the proof of Theorem 9.1 that the infimum is indeed attained). Then the unique equilibrium measure μ_α for the function $q(\alpha)(a - \alpha b)$ satisfies

$$\mathcal{F}(\alpha) = h_{\mu_\alpha}(\Phi) \quad \text{and} \quad \frac{\int_\Lambda a\, d\mu_\alpha}{\int_\Lambda b\, d\mu_\alpha} = \alpha.$$

Using similar arguments to those in [14] in the case of discrete time, one can extend Theorem 9.1 to the case when the entropy is upper semicontinuous, for continuous functions with unique equilibrium measures. For example, for a locally maximal hyperbolic set for a topologically mixing C^1 flow the entropy is upper semicontinuous and any continuous function with bounded variation has a unique equilibrium measure. We recall that a continuous function $a\colon X \to \mathbb{R}$ is said to have *bounded variation* if there exist $\varepsilon > 0$ and $\kappa > 0$ such that

$$\left| \int_0^t a(\varphi_s(x))\, ds - \int_0^t a(\varphi_s(y))\, ds \right| < \kappa$$

whenever $d(\varphi_s(x), \varphi_s(y)) < \varepsilon$ for every $s \in [0, t]$.

9.2 Analyticity of the Spectrum

Let Φ be a C^1 flow with a locally maximal hyperbolic set Λ. Given functions $a, b\colon \Lambda \to \mathbb{R}$ as in Theorem 9.1, the following result shows that when a is not Φ-cohomologous to a multiple of b the spectrum $\mathcal{F}$ in (9.1) is analytic.

Theorem 9.2 ([15]) *Let Φ be a C^1 flow with a locally maximal hyperbolic set Λ such that $\Phi|\Lambda$ is topologically mixing and let $a, b\colon \Lambda \to \mathbb{R}$ be Hölder continuous functions with $b > 0$. Then the following properties hold:*

1. *if a is Φ-cohomologous to cb in Λ for some $c \in \mathbb{R}$, then $\underline{\alpha} = \overline{\alpha} = c$ and $K_c = \Lambda$;*

2. *if a is not Φ-cohomologous to a multiple of b in Λ, then $\underline{\alpha} < \overline{\alpha}$ and the function $\mathcal{F}$ is analytic in the interval $(\underline{\alpha}, \overline{\alpha})$.*

Proof Let us assume that there exists a constant c such that a is Φ-cohomologous to cb in Λ. We have

$$\begin{aligned}
&\left|\int_0^t a(\varphi_\tau(x))\,d\tau - c\int_0^t b(\varphi_\tau(x))\,d\tau\right| \\
&= \lim_{s\to 0}\frac{1}{s}\left|\int_s^{s+t} q(\varphi_\tau(x))\,d\tau - \int_0^t q(\varphi_\tau(x))\,d\tau\right| \\
&= \lim_{s\to 0}\frac{1}{s}\left|\int_t^{s+t} q(\varphi_\tau(x))\,d\tau - \int_0^s q(\varphi_\tau(x))\,d\tau\right| \\
&\le 2\sup|q|,
\end{aligned} \tag{9.9}$$

and hence,

$$\left|\frac{\int_0^t a(\varphi_\tau(x))\,d\tau}{\int_0^t b(\varphi_\tau(x))\,d\tau} - c\right| \le \frac{2\sup|q|}{t\inf|b|}$$

for $x \in \Lambda$ and $t > 0$. Therefore, $K_c = \Lambda$. Moreover, by (9.9), for each $\mu \in \mathcal{M}$ we have

$$\begin{aligned}
0 &= \int_\Lambda \lim_{t\to\infty}\left(\frac{1}{t}\int_0^t a(\varphi_\tau(x))\,d\tau - \frac{c}{t}\int_0^t b(\varphi_\tau(x))\,d\tau\right)d\mu(x) \\
&= \int_\Lambda a\,d\mu - c\int_\Lambda b\,d\mu
\end{aligned}$$

and $\underline{\alpha} = \overline{\alpha} = c$. This establishes the first property.

For the second property, let us assume that a is not Φ-cohomologous to a multiple of b. If $\underline{\alpha} = \overline{\alpha} = c$, then the function

$$\mu \mapsto \int_\Lambda a\,d\mu - c\int_\Lambda b\,d\mu$$

is identically zero. In particular, if μ is the invariant measure supported on the periodic orbit of a point $x = \varphi_T(x)$, then

$$\frac{1}{T}\int_0^T a(\varphi_s(x))\,ds = c\frac{1}{T}\int_0^T b(\varphi_s(x))\,ds.$$

By Livschitz's theorem (see Theorem 19.2.4 in [62]), we conclude that the functions a and cb are Φ-cohomologous. This contradiction implies that $\underline{\alpha} < \overline{\alpha}$.

Now we establish the analyticity of the spectrum.

Lemma 9.1 *If for each $\alpha \in \mathbb{R}$ the function $a - \alpha b$ is not Φ-cohomologous to a constant, then the spectrum $\mathcal{F}$ is analytic in the interval $(\underline{\alpha}, \overline{\alpha})$.*

Proof of the lemma Take $\alpha \in (\underline{\alpha}, \overline{\alpha})$ and let

$$F(q, \alpha) = P_\Phi(qa - q\alpha b).$$

By Theorem 9.1, the number $\mathcal{F}(\alpha)$ coincides with $\min_{q\in\mathbb{R}} F(q, \alpha)$. Moreover, the function F is analytic in both variables. We want to apply the Implicit function theorem to show that the minimum is attained at a point $q = q(\alpha)$ depending analytically on α.

We have

$$\partial_q F(q, \alpha) = \int_\Lambda (a - \alpha b)\, d\nu_{q,\alpha},$$

where $\nu_{q,\alpha}$ is the equilibrium measure for $qa - q\alpha b$. By Theorem 9.1, there exists a $q = q(\alpha) \in \mathbb{R}$ at which the function $q \mapsto P_\Phi(qa - q\alpha b)$ attains a minimum. Thus, we have

$$\partial_q F(q(\alpha), \alpha) = 0.$$

Since $a - \alpha b$ is not Φ-cohomologous to a constant, the function $q \mapsto F(q, \alpha)$ is strictly convex. Hence, $q = q(\alpha)$ is the unique real number satisfying $\partial_q F(q, \alpha) = 0$. Again since $a - \alpha b$ is not Φ-cohomologous to a constant, the derivative $\partial_q^2 F$ does not vanish (see [92]). Thus, it follows from the Implicit function theorem that the function $\alpha \mapsto q(\alpha)$ is analytic. This completes the proof of the lemma. ☐

In view of Lemma 9.1, it remains to consider the case when there exist $c, d \in \mathbb{R}$ with $d \neq 0$ and a bounded measurable function $q\colon \Lambda \to \mathbb{R}$ such that

$$a(x) - cb(x) = d + \lim_{t\to 0} \frac{q(\varphi_t(x)) - q(x)}{t} \tag{9.10}$$

for $x \in \Lambda$. One can easily verify that

$$x \in K_\alpha(a, b) \quad \text{if and only if} \quad x \in K_{d/(\alpha-c)}(b, 1).$$

Moreover, it follows from (9.9) and (9.10) that

$$\left| \int_0^t a(\varphi_\tau(x))\, d\tau - c \int_0^t b(\varphi_\tau(x))\, d\tau - dt \right| \le 2 \sup|q|.$$

Since $b > 0$ and $d \neq 0$, we conclude that $c \neq \alpha$ for every $\alpha \in \mathbb{R}$ with $K_\alpha(a, b) \neq \varnothing$. Hence, the function $\alpha \mapsto d/(\alpha - c)$ is analytic in $(\underline{\alpha}, \overline{\alpha})$.

Now we observe that b is not Φ-cohomologous to a constant $\rho \in \mathbb{R}$. Otherwise the function a would be Φ-cohomologous to $cb + d = (c + d/\rho)b$, which yields a contradiction. Hence, one can apply Lemma 9.1 to the pair of functions $(b, 1)$ to conclude that the spectrum $\mathcal{F}^{(b,1)}$ is analytic in the nonempty interval $(\underline{\kappa}, \overline{\kappa})$, where

$$\underline{\kappa} = \inf\left\{\int_\Lambda b\, d\mu : \mu \in \mathcal{M}\right\} \quad \text{and} \quad \overline{\kappa} = \sup\left\{\int_\Lambda b\, d\mu : \mu \in \mathcal{M}\right\}.$$

Since $b > 0$, we have $\underline{\kappa} > 0$. The spectrum $\mathcal{F}^{(a,b)}$ is the composition of the analytic functions $\alpha \mapsto d/(\alpha - c)$ and $\mathcal{F}^{(b,1)}$, and thus it is also analytic. Moreover,

$$(\underline{\alpha}, \overline{\alpha}) = \begin{cases} (c + d/\overline{\kappa}, c + d/\underline{\kappa}) & \text{when } d > 0, \\ (c + d/\underline{\kappa}, c + d/\overline{\kappa}) & \text{when } d < 0. \end{cases}$$

This completes the proof of the theorem. □

In the special case when $b = 1$ the statement in Theorem 9.2 was first established in [12] (using a different method).

Now we show that most Hölder continuous functions satisfy the second alternative in Theorem 9.2. Let $C^\gamma(\Lambda)$ be the space of Hölder continuous functions in Λ with Hölder exponent $\gamma \in (0, 1)$ equipped with the norm in (8.25). We denote by $C^\gamma_+(\Lambda)$ the set of all positive functions in $C^\gamma(\Lambda)$.

Theorem 9.3 ([15]) *Let Φ be a C^1 flow with a locally maximal hyperbolic set Λ such that $\Phi|\Lambda$ is topologically transitive. For each $\gamma \in (0, 1)$, the set of all functions $(a, b) \in C^\gamma(\Lambda) \times C^\gamma_+(\Lambda)$ such that a is not Φ-cohomologous to a multiple of b is open and dense in $C^\gamma(\Lambda) \times C^\gamma_+(\Lambda)$.*

Proof Let $H = C^\gamma(\Lambda) \times C^\gamma_+(\Lambda)$. Let also $G \subset H$ be the set of all pairs $(a, b) \in H$ such that a is not Φ-cohomologous to a multiple of b. Take $(a, b) \in H \setminus G$ and let Γ_i be distinct periodic orbits of points $x_i = \varphi_{T_i}(x_i)$ for $i = 0, 1$. We write

$$\langle g \rangle_i = \frac{1}{T_i} \int_0^{T_i} g(\varphi_t(x_i))\, dt$$

for each continuous function $g \colon \Lambda \to \mathbb{R}$ and $i = 0, 1$. Consider a function $h \in C^\gamma(\Lambda)$ such that $h|\Gamma_0 = \langle b \rangle_0$ and $h|\Gamma_1 = \langle b \rangle_1 + 1$. This is always possible because Γ_0 and Γ_1 are closed and disjoint. Now we consider the pair of functions

$$(\tilde{a}, \tilde{b}) = (a, b) + (\varepsilon h, 0) \in H$$

for some constant $\varepsilon > 0$. For each $\tilde{c} \in \mathbb{R}$, we have

$$\tilde{a} - \tilde{c}\tilde{b} = a - cb + (c - \tilde{c})b + \varepsilon h.$$

Thus, if $\tilde{a} - \tilde{c}\tilde{b}$ is Φ-cohomologous to zero, then

$$0 = \langle \tilde{a} - \tilde{c}\tilde{b} \rangle_0 = (c - \tilde{c} + \varepsilon)\langle b \rangle_0$$

and

$$0 = \langle \tilde{a} - \tilde{c}\tilde{b} \rangle_1 = (c - \tilde{c} + \varepsilon)\langle b \rangle_1 + \varepsilon.$$

Since $\langle b \rangle_0 \geq \min b > 0$, we obtain $c - \tilde{c} + \varepsilon = 0$. But this is impossible, in view of the second identity. This contradiction implies that $(\tilde{a}, \tilde{b}) \in G$. Since ε is arbitrary, the pair of functions (a, b) can be arbitrarily approximated in H by pairs in G, and thus G is dense in H.

Now we show that G is open. Let $(a, b) \in G$. Since $b > 0$, there exists a unique $c = c(a, b) \in \mathbb{R}$ such that $P_\Phi(a - cb) = P_\Phi(0)$. By Livschitz's theorem, there also exists a periodic orbit Γ_0 such that $\langle a - cb\rangle_0 \neq 0$. Take $\varepsilon \in (0, \min b/2)$ and $(\tilde{a}, \tilde{b}) \in H$ such that

$$\|a - \tilde{a}\|_\gamma + \|b - \tilde{b}\|_\gamma < \varepsilon.$$

We have

$$|P_\Phi(\tilde{a} - c\tilde{b}) - P_\Phi(0)| \leq \|\tilde{a} - a - c(\tilde{b} - b)\|_\gamma < (1 + |c|)\varepsilon. \tag{9.11}$$

Now let $\tilde{c} \in \mathbb{R}$ be the unique real number such that $P_\Phi(\tilde{a} - \tilde{c}\tilde{b}) = P_\Phi(0)$. We observe that if $\tilde{a}$ is Φ-cohomologous to a multiple of $\tilde{b}$, then $\tilde{a}$ is Φ-cohomologous to $\tilde{c}\tilde{b}$ and to no other multiple of $\tilde{b}$. Since $\tilde{b} > \min b/2 > 0$, it follows from (9.11) that

$$|c - \tilde{c}| \leq \frac{1}{\min \tilde{b}} |P_\Phi(\tilde{a} - \tilde{c}\tilde{b}) - P_\Phi(\tilde{a} - c\tilde{b})| < \frac{2(1 + |c|)\varepsilon}{\min b}.$$

Therefore,

$$\begin{aligned}
|\langle \tilde{a} - \tilde{c}\tilde{b}\rangle_0| &\geq |\langle a - cb\rangle_0| - |\langle \tilde{a} - a - (\tilde{c}\tilde{b} - cb)\rangle_0| \\
&\geq |\langle a - cb\rangle_0| - \|\tilde{a} - a\|_\gamma - |\tilde{c} - c| \cdot \|\tilde{b}\|_\gamma - |c| \cdot \|\tilde{b} - b\|_\gamma \\
&\geq |\langle a - cb\rangle_0| - \left(1 + \frac{2(1 + |c|)(\|b\|_\gamma + \varepsilon)}{\min b} + |c|\right)\varepsilon > 0,
\end{aligned}$$

provided that ε is sufficiently small (possibly depending on a and b). This implies that $\tilde{a}$ is not Φ-cohomologous to $\tilde{c}\tilde{b}$. Hence, the ball of radius ε centered at (a, b) is contained in G. This shows that the set G is open. □

Combining Theorems 9.1, 9.2 and 9.3 we readily obtain the following result, whose formulation has the advantage of not using the notion of cohomology.

Theorem 9.4 ([15]) *Let Φ be a C^1 flow with a locally maximal hyperbolic set Λ such that $\Phi|\Lambda$ is topologically mixing. Given $\gamma \in (0, 1)$, for $(a, b) \in C^\gamma(\Lambda) \times C^\gamma_+(\Lambda)$ in an open and dense set, the entropy spectrum $\mathcal{F}$ is analytic in the nonempty interval $(\underline{\alpha}, \overline{\alpha})$ and satisfies identities* (9.2) *and* (9.3) *for $\alpha \in (\underline{\alpha}, \overline{\alpha})$.*

9.3 Examples

This section describes some applications of Theorems 9.1 and 9.2 to various spectra. In particular, we consider multifractal spectra for the local entropies, multifractal spectra for the Lyapunov exponents, and the particular case of suspension flows.

9.3.1 Multifractal Spectra for the Local Entropies

Let Φ be a continuous flow, let Λ be a Φ-invariant set, and let ν be a Φ-invariant probability measure on Λ.

Definition 9.2 For each $x \in \Lambda$, we define the *lower* and *upper* ν*-local entropies of* Φ *at* x respectively by

$$\underline{h}_\nu(\Phi, x) = \lim_{\varepsilon \to 0} \liminf_{t \to \infty} -\frac{1}{t} \log \nu(B(x, t, \varepsilon))$$

and

$$\overline{h}_\nu(\Phi, x) = \lim_{\varepsilon \to 0} \limsup_{t \to \infty} -\frac{1}{t} \log \nu(B(x, t, \varepsilon)),$$

with $B(x, t, \varepsilon)$ as in (4.1).

Whenever $\underline{h}_\nu(\Phi, x) = \overline{h}_\nu(\Phi, x)$, the common value is denoted by $h_\nu(\Phi, x)$ and is called the ν*-local entropy of* Φ *at* x. By the Shannon–McMillan–Breiman theorem, the ν-local entropy of Φ is well defined ν-almost everywhere. In addition, if ν is ergodic, then $h_\nu(\Phi, x) = h_\nu(\Phi)$ for ν-almost every $x \in \Lambda$.

Definition 9.3 The *entropy spectrum for the local entropies* of ν is defined by

$$\mathcal{H}(\alpha) = h(\Phi | K_\alpha^h),$$

where

$$K_\alpha^h = \left\{ x \in \Lambda : \underline{h}_\nu(\Phi, x) = \overline{h}_\nu(\Phi, x) = \alpha \right\}.$$

Now let Λ be a locally maximal hyperbolic set for Φ. In this case we have

$$K_\alpha^h = \left\{ x \in \Lambda : \lim_{t \to \infty} -\frac{1}{t} \log \nu(B(x, t, \varepsilon)) = \alpha \right\}$$

for any sufficiently small $\varepsilon > 0$. Moreover, there exists a unique measure m_E of maximal entropy, that is, a Φ-invariant probability measure on Λ such that $h_{m_E}(\Phi) = h(\Phi)$. We write

$$\underline{\alpha}^h = \inf \left\{ -\int_\Lambda a \, d\mu : \mu \in \mathcal{M} \right\}$$

and

$$\overline{\alpha}^h = \sup \left\{ -\int_\Lambda a \, d\mu : \mu \in \mathcal{M} \right\}.$$

The following result gives a conditional variational principle for the spectrum $\mathcal{H}$.

Theorem 9.5 ([15]) *Let Φ be a C^1 flow with a locally maximal hyperbolic set Λ such that $\Phi|\Lambda$ is topologically mixing and let ν be an equilibrium measure for a Hölder continuous function $a\colon \Lambda \to \mathbb{R}$ such that $P_\Phi(a) = 0$. Then the following properties hold*:

1. *if $\alpha \notin [\underline{\alpha}^h, \overline{\alpha}^h]$, then $K_\alpha^h = \varnothing$;*
2. *if $\alpha \in (\underline{\alpha}^h, \overline{\alpha}^h)$, then $K_\alpha^h \neq \varnothing$ and*

$$\begin{aligned}\mathcal{H}(\alpha) &= \max\left\{h_\mu(\Phi) : -\int_\Lambda a\,d\mu = \alpha \text{ and } \mu \in \mathcal{M}\right\}\\ &= \min\{P_\Phi(qa) + q\alpha : q \in \mathbb{R}\};\end{aligned}$$

3. *if $\nu = m_E$, that is, a is Φ-cohomologous to zero, then $\underline{\alpha}^h = \overline{\alpha}^h = c$ and $K_c^h = \Lambda$;*
4. *if $\nu \neq m_E$, that is, a is not Φ-cohomologous to zero, then $\underline{\alpha}^h < \overline{\alpha}^h$ and the function $\mathcal{H}$ is analytic in the interval $(\underline{\alpha}^h, \overline{\alpha}^h)$.*

Proof The result follows from Theorems 9.1 and 9.2 taking $b = -1$.

9.3.2 Multifractal Spectra for the Lyapunov Exponents

In this section we consider the multifractal spectrum for the Lyapunov exponents. Let Φ be a C^1 flow with a locally maximal hyperbolic set Λ such that $\Phi|\Lambda$ is conformal (see Definition 5.1).

Let Z_s and Z_u be respectively the sets of points $x \in \Lambda$ such that each of the limits

$$\lambda_s(x) = \lim_{t\to+\infty} \frac{1}{t} \log\|d_x\varphi_t|E^s(x)\| \quad \text{and} \quad \lambda_u(x) = \lim_{t\to+\infty} \frac{1}{t} \log\|d_x\varphi_t|E^u(x)\|$$

exists. As in Sect. 6.1, for any Φ-invariant probability measure ν in Λ we have

$$\nu(\Lambda \setminus Z_s) = \nu(\Lambda \setminus Z_u) = 0.$$

Definition 9.4 The *stable* and *unstable entropy spectra for the Lyapunov exponents* are defined respectively by

$$\mathcal{L}_s(\alpha) = h(\Phi|K_\alpha^s) \quad \text{and} \quad \mathcal{L}_u(\alpha) = h(\Phi|K_\alpha^u),$$

where

$$K_\alpha^s = \{x \in Z_s : \lambda_s(x) = \alpha\} \quad \text{and} \quad K_\alpha^u = \{x \in Z_u : \lambda_u(x) = \alpha\}.$$

The following result gives a conditional variational principle for the spectrum $\mathcal{L}_s$. We write

$$\underline{\alpha}_s = \inf\left\{\int_\Lambda \zeta_s\,d\mu : \mu \in \mathcal{M}\right\}$$

and

$$\overline{\alpha}_s = \sup\left\{\int_\Lambda \zeta_s \, d\mu : \mu \in \mathcal{M}\right\}.$$

Theorem 9.6 [15] *Let Φ be a $C^{1+\delta}$ flow with a locally maximal hyperbolic set Λ such that $\Phi|\Lambda$ is conformal and topologically mixing. Then the following properties hold*:

1. *if $\alpha \notin [\underline{\alpha}_s, \overline{\alpha}_s]$, then $K_\alpha^s = \varnothing$;*
2. *if $\alpha \in (\underline{\alpha}_s, \overline{\alpha}_s)$, then $K_\alpha^s \neq \varnothing$ and*

$$\begin{aligned}\mathcal{L}_s(\alpha) &= \max\left\{h_\mu(\Phi) : \int_\Lambda \zeta_s \, d\mu = \alpha \text{ and } \mu \in \mathcal{M}\right\} \\ &= \min\{P_\Phi(q\zeta_s) - q\alpha : q \in \mathbb{R}\};\end{aligned}$$

3. *if ζ_s is Φ-cohomologous to zero, then $\underline{\alpha}_s = \overline{\alpha}_s = c$ and $K_c^s = \Lambda$;*
4. *if ζ_s is not Φ-cohomologous to zero, then $\underline{\alpha}_s < \overline{\alpha}_s$ and the function $\mathcal{L}_s$ is analytic in the interval $(\underline{\alpha}_s, \overline{\alpha}_s)$.*

Proof Since the stable and unstable distributions $x \mapsto E^s(x)$ and $x \mapsto E^u(x)$ are Hölder continuous and the flow Φ is of class $C^{1+\delta}$, the functions ζ_s and ζ_u are also Hölder continuous in Λ. Hence, the result follows from Theorems 9.1 and 9.2 taking $a = \zeta_s$ and $b = 1$. □

In [82], Pesin and Sadovskaya obtained a multifractal analysis of the spectrum $\mathcal{L}_s$. One can also formulate corresponding statements for the spectrum $\mathcal{L}_u$.

9.3.3 Suspension Flows

Let Ψ be a suspension flow in Y, over a homeomorphism $T : X \to X$ of the compact metric space X, and let μ be a T-invariant probability measure on X.

Given continuous functions $a, b : Y \to \mathbb{R}$ with $b > 0$ and $\alpha \in \mathbb{R}$, let

$$K_\alpha = \left\{x \in Y : \lim_{t\to\infty} \frac{\int_0^t a(\psi_s(x))\, ds}{\int_0^t b(\psi_s(x))\, ds} = \alpha\right\}$$

and consider again the spectrum $\mathcal{F}$ in (9.1). It follows from Proposition 7.2 that the set K_α is composed of the points $(x, s) \in Y$ such that

$$\lim_{m\to\infty} \frac{\sum_{i=0}^m I_a(T^i(x))}{\sum_{i=0}^m I_b(T^i(x))} = \alpha$$

and $s \in [0, \tau(x)]$. Let also

$$\underline{\alpha} = \inf\left\{\frac{\int_Y a\,d\nu}{\int_Y b\,d\nu} : \nu \in \mathcal{M}_\Psi\right\} = \inf\left\{\frac{\int_X I_a\,d\mu}{\int_X I_b\,d\mu} : \mu \in \mathcal{M}_T\right\}$$

and

$$\overline{\alpha} = \sup\left\{\frac{\int_Y a\,d\nu}{\int_Y b\,d\nu} : \nu \in \mathcal{M}_\Psi\right\} = \inf\left\{\frac{\int_X I_a\,d\mu}{\int_X I_b\,d\mu} : \mu \in \mathcal{M}_T\right\},$$

where $\mathcal{M}_\Psi$ (respectively $\mathcal{M}_T$) is the set of all Ψ-invariant probability measures on Y (respectively of all T-invariant probability measures on X).

The following result gives a conditional variational principle for the spectrum $\mathcal{F}$ in the special case when T is a topological Markov chain.

Theorem 9.7 ([15]) *Let Ψ be a suspension flow over a topologically mixing two-sided topological Markov chain and let $a, b\colon Y \to \mathbb{R}$ be Hölder continuous functions with $b > 0$. Then the following properties hold*:

1. *if $\alpha \notin [\underline{\alpha}, \overline{\alpha}]$, then $K_\alpha = \varnothing$;*
2. *if $\alpha \in (\underline{\alpha}, \overline{\alpha})$, then $K_\alpha \neq \varnothing$ and*

$$\begin{aligned}\mathcal{F}(\alpha) &= \max\left\{\frac{h_\mu(T)}{\int_X \tau\,d\mu} : \frac{\int_X I_a\,d\mu}{\int_X I_b\,d\mu} = \alpha \text{ and } \mu \in \mathcal{M}_T\right\}\\ &= \min\left\{\sup_{\mu\in\mathcal{M}_T} \frac{h_\mu(T) + \int_X I_{qa-q\alpha b}\,d\mu}{\int_X \tau\,d\mu} : q \in \mathbb{R}\right\};\end{aligned}$$

3. *if a is Ψ-cohomologous to cb for some $c \in \mathbb{R}$, that is, I_a is T-cohomologous to cI_b for some $c \in \mathbb{R}$, then $\underline{\alpha} = \overline{\alpha} = c$ and $K_c = \Lambda$;*
4. *if a is not Ψ-cohomologous to a multiple of b, that is, I_a is not T-cohomologous to a multiple of I_b, then $\underline{\alpha} < \overline{\alpha}$ and the function $\mathcal{F}$ is analytic in the interval $(\underline{\alpha}, \overline{\alpha})$.*

Proof It follows from (4.9) that

$$\frac{\int_Y a\,d\nu}{\int_Y b\,d\nu} = \frac{\int_X I_a\,d\mu}{\int_X I_b\,d\mu}. \tag{9.12}$$

On the other hand, by Abramov's entropy formula, we have

$$h_\nu(\Psi) = \frac{h_\mu(T)}{\int_X \tau\,d\mu}. \tag{9.13}$$

By (9.12) and (9.13), using similar arguments to those in the proof of Theorem 9.1 we obtain the first and second properties in the theorem. The remaining properties follow from Theorem 2.1, using similar arguments to those in the proof of Theorem 9.2. □

9.4 Multidimensional Spectra

In this section we describe a multidimensional version of the conditional variational principle in Theorem 9.1. This can be seen as a motivation for Chap. 10 where we establish much more general results. Instead of considering Birkhoff averages (or ratios of Birkhoff averages) we consider vectors of ratios of Birkhoff averages.

Let Φ be a flow and let Λ be a Φ-invariant set. Let also $a_1, \dots, a_d \colon \Lambda \to \mathbb{R}$ and $b_1, \dots, b_d \colon \Lambda \to \mathbb{R}$ be continuous functions with $b_i > 0$ for $i = 1, \dots, d$. We write

$$A = \left\{ \left(\frac{\int_\Lambda a_1\, d\mu}{\int_\Lambda b_1\, d\mu}, \dots, \frac{\int_\Lambda a_d\, d\mu}{\int_\Lambda b_d\, d\mu} \right) : \mu \in \mathcal{M} \right\},$$

where $\mathcal{M}$ is the set of all Φ-invariant probability measures on Λ, and we define

$$K_\alpha = \left\{ x \in \Lambda : \lim_{t\to\infty} \left(\frac{\int_0^t a_1(\varphi_s(x))\, ds}{\int_0^t b_1(\varphi_s(x))\, ds}, \dots, \frac{\int_0^t a_d(\varphi_s(x))\, ds}{\int_0^t b_d(\varphi_s(x))\, ds} \right) = \alpha \right\}$$

for each $\alpha = (\alpha_1, \dots, \alpha_d) \in \mathbb{R}^d$.

Definition 9.5 The function $\mathcal{F} = \mathcal{F}^{(a,b)}$ defined by

$$\mathcal{F}(\alpha) = h(\Phi | K_\alpha)$$

is called the *entropy spectrum* for the pair $(a, b) = (a_1, \dots, a_d, b_1, \dots, b_d)$.

The following result gives a conditional variational principle for the spectrum $\mathcal{F}$. It is a multidimensional version of Theorem 9.1.

Theorem 9.8 ([15]) *Let Φ be a C^1 flow with a locally maximal hyperbolic set Λ such that $\Phi|\Lambda$ is topologically mixing and let $a_1, \dots, a_d$ and $b_1, \dots, b_d$ be Hölder continuous functions with $b_i > 0$ for each $i = 1, \dots, d$. Then the following properties hold*:

1. *if $\alpha \notin A$, then $K_\alpha = \varnothing$;*
2. *if $\alpha \in \operatorname{int} A$, then $K_\alpha \neq \varnothing$ and*

$$\mathcal{F}(\alpha) = \max \left\{ h_\mu(\Phi) : \left(\frac{\int_\Lambda a_1\, d\mu}{\int_\Lambda b_1\, d\mu}, \dots, \frac{\int_\Lambda a_d\, d\mu}{\int_\Lambda b_d\, d\mu} \right) = \alpha \text{ and } \mu \in \mathcal{M} \right\}$$

$$= \min \left\{ P_\Phi \left(\sum_{i=1}^d (q_i a_i - q_i \alpha_i b_i) \right) : (q_1, \dots, q_d) \in \mathbb{R}^d \right\}.$$

Proof The first property can be obtained in a similar manner to that in the proof of the first property in Theorem 9.1.

For the second property, we briefly describe the changes that are required in the proof of Theorem 9.1 when $d > 1$. For each $\alpha = (\alpha_1, \dots, \alpha_d) \in \operatorname{int} A$ and $q = (q_1, \dots, q_d) \in \mathbb{R}^d \setminus \{0\}$, take measures ν_-^q and μ_+^q such that

$$\sum_{i=1}^{d} q_i \left(\int_\Lambda a_i \, d\nu_-^q - \alpha_i \int_\Lambda b_i \, d\nu_-^q \right) < 0 < \sum_{i=1}^{d} q_i \left(\int_\Lambda a_i \, d\nu_+^q - \alpha_i \int_\Lambda b_i \, d\nu_+^q \right).$$

These play the role of the measures ν_- and ν_+ in the proof of Theorem 9.1 and similar arguments can be used to show that

$$\liminf_{\|q\| \to \infty} P_\Phi \left(\sum_{i=1}^{d} (q_i a_i - q_i \alpha_i b_i) \right) = +\infty,$$

where $\|\cdot\|$ is any norm in $\mathbb{R}^n$. This implies that the function

$$F \colon (q_1, \dots, q_d) \mapsto P_\Phi \left(\sum_{i=1}^{d} (q_i a_i - q_i \alpha_i b_i) \right)$$

attains its infimum at some point $q(\alpha) \in \mathbb{R}^d$, and hence $\partial_q F(q(\alpha)) = 0$. This property allows one to use essentially the same arguments as in the proof of Theorem 9.1, replacing a and b by the vectors $(a_1, \dots, a_d)$ and $(b_1, \dots, b_d)$, to obtain the desired result. □

Theorem 9.8 is a particular case of Theorem 10.1 and for this reason we have only sketched the proof. Theorem 10.1 gives a conditional variational principle for multidimensional BS-dimension spectra (Theorem 9.8 considers the particular case of multidimensional entropy spectra).

Chapter 10
Multidimensional Spectra

In this chapter we present a multidimensional multifractal analysis for hyperbolic flows. More precisely, we consider multifractal spectra associated to multidimensional parameters, obtained from computing the entropy of the level sets associated to several Birkhoff averages. These spectra exhibit several new phenomena that are absent in 1-dimensional multifractal analysis. We also consider the more general class of flows with upper semicontinuous entropy. In this chapter the multifractal analysis is obtained from a conditional variational principle for the topological entropy of the level sets.

10.1 Multifractal Analysis

In this section we consider a multidimensional multifractal spectrum for ratios of Birkhoff averages of a flow and we establish a corresponding conditional variational principle.

Let $\Phi = \{\varphi_t\}_{t\in\mathbb{R}}$ be a continuous flow in a compact metric space X. We consider vectors of functions $(A, B) \in C(X)^d \times C(X)^d$ for some $d \in \mathbb{N}$, say with components

$$A = (a_1, \dots, a_d) \quad \text{and} \quad B = (b_1, \dots, b_d),$$

with $b_i > 0$ for $i = 1, \dots, d$. We equip $\mathbb{R}^d$ with the norm $\|\alpha\| = |\alpha_1| + \cdots + |\alpha_d|$ and $C(X)^d$ with the corresponding supremum norm. For each $\alpha = (\alpha_1, \dots, \alpha_d) \in \mathbb{R}^d$, let

$$K_\alpha = K_\alpha(A, B) = \bigcap_{i=1}^{d} \left\{ x \in X : \lim_{t\to\infty} \frac{\int_0^t a_i(\varphi_s(x))ds}{\int_0^t b_i(\varphi_s(x))ds} = \alpha_i \right\}. \tag{10.1}$$

Definition 10.1 Given a continuous function $u\colon X \to \mathbb{R}^+$, the *BS-dimension spectrum* $\mathcal{F}_u\colon \mathbb{R}^d \to \mathbb{R}$ of the pair (A, B) (with respect to u and Φ) is defined by

$$\mathcal{F}_u(\alpha) = \dim_u K_\alpha(A, B). \tag{10.2}$$

L. Barreira, *Dimension Theory of Hyperbolic Flows*,
Springer Monographs in Mathematics, DOI 10.1007/978-3-319-00548-5_10,

We also consider the function $\mathcal{P} = \mathcal{P}_{A,B} \colon \mathcal{M} \to \mathbb{R}^d$ defined by

$$\mathcal{P}(\mu) = \left(\frac{\int_X a_1\, d\mu}{\int_X b_1\, d\mu}, \ldots, \frac{\int_X a_d\, d\mu}{\int_X b_d\, d\mu} \right),$$

where $\mathcal{M}$ is the set of all Φ-invariant probability measures on X. For each $\alpha = (\alpha_1, \ldots, \alpha_d)$ and $\beta = (\beta_1, \ldots, \beta_d)$ in $\mathbb{R}^d$, we write

$$\alpha * \beta = (\alpha_1\beta_1, \ldots, \alpha_d\beta_d) \in \mathbb{R}^d \quad \text{and} \quad \langle \alpha, \beta \rangle = \sum_{i=1}^d \alpha_i \beta_i \in \mathbb{R}.$$

The following result gives a conditional variational principle for the spectrum $\mathcal{F}_u$.

Theorem 10.1 ([6]) *Let Φ be a continuous flow in a compact metric space X such that the map $\mu \mapsto h_\mu(\Phi)$ is upper semicontinuous, and consider functions $(A, B) \in C(X)^d \times C(X)^d$ such that*

$$\operatorname{span}\{a_1, b_1, \ldots, a_d, b_d, u\} \subset D(X).$$

If $\alpha \in \operatorname{int} \mathcal{P}(\mathcal{M})$, then $K_\alpha \neq \varnothing$ and the following properties hold:

1.

$$\mathcal{F}_u(\alpha) = \max \left\{ \frac{h_\mu(\Phi)}{\int_X u\, d\mu} : \mu \in \mathcal{M} \text{ and } \mathcal{P}(\mu) = \alpha \right\}; \tag{10.3}$$

2.

$$\mathcal{F}_u(\alpha) = \min\{T_u(\alpha, q) : q \in \mathbb{R}^d\}, \tag{10.4}$$

where $T_u(\alpha, q)$ is the unique real number such that

$$P_\Phi\big(\langle q, A - \alpha * B \rangle - T_u(\alpha, q) u\big) = 0; \tag{10.5}$$

3. *there exists an ergodic measure $\mu_\alpha \in \mathcal{M}$ such that $\mathcal{P}(\mu_\alpha) = \alpha$, $\mu_\alpha(K_\alpha) = 1$ and* $\dim_u \mu_\alpha = \mathcal{F}_u(\alpha)$.

Moreover, if $\alpha \notin \mathcal{P}(\mathcal{M})$, then $K_\alpha = \varnothing$.

Proof The proof follows arguments of Barreira, Saussol and Schmeling in [16] in the case of discrete time. We use the notation $\mu(\psi) = \int_X \psi\, d\mu$.

Take $\alpha \in \mathbb{R}^d$ such that $K_\alpha \neq \varnothing$. Given $x \in K_\alpha$, we define a sequence $(\mu_n)_{n \in \mathbb{N}}$ of probability measures on X by

$$\mu_n(a) = \frac{1}{n} \int_0^n a(\varphi_s(x))\, ds$$

for each $a \in C(X)$. Since $\mathcal{M}$ is compact, this sequence has at least one accumulation point $\mu \in \mathcal{M}$. Therefore,

$$\begin{aligned}\alpha &= \left(\lim_{t\to+\infty}\frac{\int_0^t a_1(\varphi_s(x))\,ds}{\int_0^t b_1(\varphi_s(x))\,ds},\dots,\lim_{t\to+\infty}\frac{\int_0^t a_d(\varphi_s(x))\,ds}{\int_0^t b_d(\varphi_s(x))\,ds}\right)\\ &= \left(\lim_{n\to+\infty}\frac{\mu_n(a_1)}{\mu_n(b_1)},\dots,\lim_{n\to+\infty}\frac{\mu_n(a_d)}{\mu_n(b_d)}\right)\\ &= \left(\frac{\mu(a_1)}{\mu(b_1)},\dots,\frac{\mu(a_d)}{\mu(b_d)}\right) = \mathcal{P}(\mu) \in \mathcal{P}(\mathcal{M}).\end{aligned}$$

Now let $\alpha \in \operatorname{int}\mathcal{P}(\mathcal{M})$. The existence of the maximum in (10.3) is a consequence of the upper semicontinuity of the map $\mu \mapsto h_\mu(\Phi)/\int_X u\,d\mu$, together with the compactness of $\mathcal{M}$ and the continuity of $\mathcal{P}$. For each $q \in \mathbb{R}^d$, let

$$\varphi_{q,\alpha} = \langle q, A - \alpha * B\rangle - \mathcal{F}_u(\alpha)u \quad\text{and}\quad F_\alpha(q) = P_\Phi(\varphi_{q,\alpha}).$$

Let also $r > 0$ be the distance from α to $\mathbb{R}^d \setminus \mathcal{P}(\mathcal{M})$ and take q such that

$$\|q\| \geq \frac{\dim_u X \cdot \sup u + F_\alpha(0)}{r \min_i \inf b_i} = R.$$

For each $\lambda \in (0,1)$ and $\beta = (\beta_1,\dots,\beta_d) \in \mathbb{R}^d$ with

$$\beta_i = \alpha_i + \frac{1}{d}\lambda r \operatorname{sgn} q_i,$$

we have

$$\|\beta - \alpha\| = \sum_{i=1}^d |\beta_i - \alpha_i| = \sum_{i=1}^d \frac{1}{d}\lambda r \operatorname{sgn}|q_i| = \lambda r < r.$$

Hence, $\beta \in \mathcal{P}(\mathcal{M})$ and there exists a $\mu \subset \mathcal{M}$ such that $\mu(A - \beta * B) = 0$. Therefore,

$$\begin{aligned}\big\langle q, \mu(A - \alpha * B)\big\rangle &= \big\langle q, \mu((\beta - \alpha) * B)\big\rangle\\ &= \sum_{i=1}^d q_i \mu((\beta_i - \alpha_i) * b_i)\\ &= \sum_{i=1}^d \frac{1}{d}\lambda r q_i \operatorname{sgn} q_i \int_X b_i\,d\mu\\ &= \frac{1}{d}\lambda r \sum_{i=1}^d |q_i| \int_X b_i\,d\mu\\ &\geq \lambda r\ \|q\| \min_i \inf b_i.\end{aligned}$$

Since $h_\mu(\Phi) \ge 0$, it follows from Proposition 4.2 that

$$\begin{aligned}
F_\alpha(q) &\ge h_\mu(\Phi) + \mu(\varphi_{q,\alpha}) \\
&= h_\mu(\Phi) + \langle q, \mu(A - \alpha * B)\rangle - \mathcal{F}_u(\alpha)\mu(u) \\
&\ge \|q\|\lambda r \min_i \inf b_i - \dim_u X \cdot \sup u \\
&\ge \lambda\left[\dim_u X \cdot \sup u + F_\alpha(0)\right] - \dim_u X \cdot \sup u.
\end{aligned}$$

Letting $\lambda \to 1$, we obtain $F_\alpha(q) \ge F_\alpha(0)$ for every $q \in \mathbb{R}^d$ such that $\|q\| \ge R$. By Proposition 4.3, the function F is of class C^1 and hence it reaches a minimum at a point $q = q(\alpha)$ with $\|q(\alpha)\| \le R$. In particular, $\partial_q F_\alpha(q(\alpha)) = 0$. By (4.6), we have

$$\mu_\alpha(A - \alpha * B) = \partial_q F_\alpha(q(\alpha)) = 0,$$

where μ_α is the equilibrium measure for $\varphi_{q,\alpha}$. This shows that $\mathcal{P}(\mu_\alpha) = \alpha$. Moreover,

$$F_\alpha(q(\alpha)) = h_{\mu_\alpha}(\Phi) - \mathcal{F}_u(\alpha)\int_X u\,d\mu_\alpha. \tag{10.6}$$

Now take $x \in K_\alpha$. For $i = 1, \ldots, d$, we have

$$\lim_{t\to\infty} \frac{\int_0^t a_i(\varphi_s(x))\,ds}{\int_0^t b_i(\varphi_s(x))\,ds} = \alpha_i.$$

Since $b_i > 0$, for each $\delta > 0$ there exists a $\tau > 0$ such that

$$\left|\frac{\int_0^t a_i(\varphi_s(x))\,ds}{\int_0^t b_i(\varphi_s(x))\,ds} - \alpha_i\right| < \frac{\delta}{dM}$$

for all $t > \tau$, where

$$M = \max_{i\in\{1,\ldots,d\}} \max_{x\in X} b_i(x).$$

We define

$$A_t(x) = \int_0^t A(\varphi_s(x))\,ds \quad \text{and} \quad B_t(x) = \int_0^t B(\varphi_s(x))\,ds, \tag{10.7}$$

and we let

$$L_{\delta,\tau} = \left\{x \in X : \|A_t(x) - \alpha * B_t(x)\| < \delta t \text{ for all } t \ge \tau\right\}.$$

Then

$$\|A_t(x) - \alpha * B_t(x)\| = \sum_{i=1}^{d} \left| \int_0^t a_i(\varphi_s(x))\,ds - \alpha_i \int_0^t b_i(\varphi_s(x))\,ds \right|$$
$$< \frac{\delta}{dM} \sum_{i=1}^{d} \int_0^t b_i(\varphi_s(x))\,ds < \delta t,$$

and hence,

$$x \in L_{\delta,\tau} \subseteq \bigcup_{\tau \in \mathbb{R}} L_{\delta,\tau}$$

for $\delta > 0$. Therefore,

$$K_\alpha \subseteq \bigcap_{\delta>0} \bigcup_{\tau \in \mathbb{R}} L_{\delta,\tau}.$$

Since X is compact, each function a_i is uniformly continuous. Hence, there exists an $\varepsilon > 0$ such that if $(x,t) \in X \times [0,\infty)$ and $y, z \in B(x,t,\varepsilon)$, and thus also $d(\varphi_s(y), \varphi_s(z)) < 2\varepsilon$, then

$$|a_i(\varphi_s(z)) - a_i(\varphi_s(y))| < \delta/d \quad \text{whenever} \quad 0 \le s \le t.$$

Let

$$A(x,t,\varepsilon) = (a_1(x,t,\varepsilon), \dots, a_d(x,t,\varepsilon))$$

and take $y \in B(x,t,\varepsilon)$. We obtain

$$\|A(x,t,\varepsilon) - A_t(y)\| = \sum_{i=1}^{d} \left| a_i(x,t,\varepsilon) - \int_0^t a_i(\varphi_s(y))\,ds \right|$$
$$\le d \sup \left\{ \int_0^t |a_i(\varphi_s(z)) - a_i(\varphi_s(y))|\,ds : z \in B(x,t,\varepsilon) \right\}$$
$$\le d \sup \left\{ \int_0^t \frac{\delta}{d}\,ds : z \in B(x,t,\varepsilon) \right\} \le \delta t,$$

and analogously,

$$\|B(x,t,\varepsilon) - B_t(y)\| \le \delta t.$$

Now take $q \in \mathbb{R}^d$. Given $(x,t) \in X \times [\tau,\infty)$ with $B(x,t,\varepsilon) \cap L_{\delta,\tau} \neq \varnothing$ and $y \in B(x,t,\varepsilon) \cap L_{\delta,\tau}$, we have

$$\begin{aligned}
-\langle q, A - \alpha * B\rangle(x,t,\varepsilon) &\le |\langle q, A - \alpha * B\rangle(x,t,\varepsilon)| \\
&\le \|q\| \cdot \|A(x,t,\varepsilon) - \alpha * B(x,t,\varepsilon)\| \\
&\le \|q\| \cdot \|A(x,t,\varepsilon) - A_t(y)\| \\
&\quad + \|q\| \cdot \|\alpha * B_t(y) - \alpha * B(x,t,\varepsilon)\| \\
&\quad + \|q\| \cdot \|A_t(y) - \alpha * B_t(y)\| \\
&\le \|q\|(\delta t + \|\alpha\|\delta t + \delta t) = c\delta t,
\end{aligned}$$

where $c = (2 + \|\alpha\|)\|q\|$. Hence,

$$\begin{aligned}
\exp\bigl(-\mathfrak{F}_u(\alpha)u(x,t,\varepsilon) - \beta t\bigr) &= \exp\bigl(\varphi_{q,\alpha}(x,t,\varepsilon) - \langle q, A - \alpha * B\rangle(x,t,\varepsilon) - \beta t\bigr) \\
&\le \exp\bigl(\varphi_{q,\alpha}(x,t,\varepsilon) - (\beta - c\delta)t\bigr)
\end{aligned}$$

for $\beta \in \mathbb{R}$. Let $T \ge \tau$ and consider a finite or countable family $\Gamma = \{(x_i,t_i)\}_{i \in I}$ such that $x_i \in X$ and $t_i \ge T$ for $i \in I$, $L_{\delta,\tau} \subset \bigcup_{i \in I} B(x_i,t_i,\varepsilon)$, and with the property that there exists no pair (x_i,t_i) such that $B(x_i,t_i,\varepsilon) \cap L_{\delta,\tau} = \varnothing$. Then

$$\sum_{(x,t)\in\Gamma} \exp(-\mathfrak{F}_u(\alpha)u(x,t,\varepsilon) - \beta t) \le \sum_{(x,t)\in\Gamma} \exp(\varphi_{q,\alpha}(x,t,\varepsilon) - (\beta - c\delta)t).$$

Taking the infimum over Γ and letting $T \to \infty$, we obtain

$$M\bigl(L_{\delta,\tau}, -\mathfrak{F}_u(\alpha)u, \beta, \varepsilon\bigr) \le M\bigl(L_{\delta,\tau}, \varphi_{q,\alpha}, \beta - c\delta, \varepsilon\bigr).$$

Letting $\varepsilon \to 0$ yields the inequality

$$P_{\Phi|L_{\delta,\tau}}(-\mathfrak{F}_u(\alpha)u) \le P_{\Phi|L_{\delta,\tau}}(\varphi_{q,\alpha}) + c\delta$$

for $\delta > 0$ and $q \in \mathbb{R}^d$. By Proposition 4.5 and the properties of the topological pressure, we have

$$\begin{aligned}
0 &= P_{\Phi|K_\alpha}(-\mathfrak{F}_u(\alpha)u) \\
&\le P_{\Phi|\bigcup_{\tau\in\mathbb{R}} L_{\delta,\tau}}(-\mathfrak{F}_u(\alpha)u) \\
&= \sup_{\tau>0} P_{\Phi|L_{\delta,\tau}}(-\mathfrak{F}_u(\alpha)u) \\
&\le P_{\Phi|L_{\delta,\tau}}(\varphi_{q,\alpha}) + c\delta \le F_\alpha(q) + c\delta
\end{aligned}$$

for $\delta > 0$ and $q \in \mathbb{R}^d$. Since δ is arbitrary, we obtain $F_\alpha(q) \ge 0$. By Proposition 4.3 and (10.6), the measure μ_α is ergodic and

$$\dim_u \mu_\alpha = \frac{h_{\mu_\alpha}(\Phi)}{\int_X u\, d\mu_\alpha} \ge \mathfrak{F}_u(\alpha).$$

On the other hand, since $\mu_\alpha(A - \alpha * B) = 0$, it follows from Birkhoff's ergodic theorem that $\mu_\alpha(K_\alpha) = 1$. This implies that

$$\begin{aligned}\mathcal{F}_u(\alpha) = \dim_u K_\alpha &= \lim_{\varepsilon\to 0} \dim_{u,\varepsilon} K_\alpha \\ &\geq \lim_{\varepsilon\to 0} \dim_{u,\varepsilon} \mu_\alpha = \dim_u \mu_\alpha,\end{aligned}$$

and hence $\dim_u \mu_\alpha = \mathcal{F}_u(\alpha)$. Therefore,

$$\begin{aligned}\min\{F_\alpha(q) : q \in \mathbb{R}^d\} = F_\alpha(q(\alpha)) &= h_{\mu_\alpha}(\Phi) - \mathcal{F}_u(\alpha) \int_X u\, d\mu_\alpha \\ &= h_{\mu_\alpha}(\Phi) - \frac{h_{\mu_\alpha}(\Phi)}{\int_X u\, d\mu_\alpha} \int_X u\, d\mu_\alpha = 0.\end{aligned}$$

Now take $\mu \in \mathcal{M}$ such that $\mathcal{P}(\mu) = \alpha$. Then $\mu(\langle q, A - \alpha * B\rangle) = 0$ and by Proposition 4.2, we have

$$\begin{aligned}0 &= \min\{F_\alpha(q) : q \in \mathbb{R}^d\} \\ &\geq \inf_{q\in\mathbb{R}^d} \bigl\{h_\mu(\Phi) + \mu\bigl(\langle q, A - \alpha * B\rangle - \mathcal{F}_u(\alpha)u\bigr)\bigr\} \\ &\geq \inf_{q\in\mathbb{R}^d} \left\{h_\mu(\Phi) - \mathcal{F}_u(\alpha) \int_X u\, d\mu\right\} \\ &= h_\mu(\Phi) - \mathcal{F}_u(\alpha) \int_X u\, d\mu.\end{aligned}$$

Therefore, $h_\mu(\Phi)/\int_X u\, d\mu \leq \mathcal{F}_u(\alpha)$, with equality when $\mu = \mu_\alpha$. This establishes properties 1 and 3 in the theorem.

Furthermore, since $F_\alpha(q(\alpha)) = 0$, we have

$$\mathcal{F}_u(\alpha) = T_u(\alpha, q(\alpha)) \geq \inf\bigl\{T_u(\alpha, q) : q \in \mathbb{R}^d\bigr\}.$$

On the other hand,

$$F_\alpha(q) \geq 0 = P_\Phi\bigl(\langle q, A - \alpha * B\rangle - T_u(\alpha, q)u\bigr),$$

and hence,

$$\mathcal{F}_u(\alpha) \leq \inf\bigl\{T_u(\alpha, q) : q \in \mathbb{R}^d\bigr\}.$$

This completes the proof of the theorem. □

As a consequence of Proposition 4.4, the conditional variational principle in Theorem 10.1 applies in particular to a topologically mixing flow on a locally maximal hyperbolic set. In this context, the statement in Theorem 10.1 was first established by Barreira and Saussol in [15] in the case of the entropy (see Theorem 9.8).

It also follows from the proof of Theorem 10.1 that μ_α can be chosen to be any equilibrium measure for the function $\langle q(\alpha), A-\alpha * B\rangle - \mathcal{F}_u(\alpha)u$, where $q(\alpha) \in \mathbb{R}^d$ is any vector such that

$$P_\Phi\big(\langle q(\alpha), A-\alpha * B\rangle - \mathcal{F}_u(\alpha)u\big) = 0.$$

We note that $q(\alpha)$ and μ_α need not be unique. The function T_u is implicitly defined by (10.5) and thus, by Proposition 4.3, the function

$$(p,\alpha,q) \mapsto P_\Phi\big(\langle q, A-\alpha * B\rangle - pu\big)$$

is of class C^1. Moreover,

$$\frac{\partial}{\partial p} P_\Phi(\langle q, A-\alpha * B\rangle - pu)\Big|_{(p,q)=(T_u(\alpha,q),q)} = -\int_X u\,d\mu_q < 0,$$

where μ_q is the equilibrium measure for $\langle q, A-\alpha * B\rangle - T_u(\alpha,q)u$. It follows from the Implicit function theorem that T_u is of class C^1 in $\mathbb{R}^d \times \mathbb{R}^d$. This implies that for each α the minimum in (10.4) is attained at a point $q \in \mathbb{R}^d$ such that $\partial_q T_u(\alpha, q) = 0$.

10.2 Finer Structure

In this section we study in greater detail the structure of a class of level sets K_α in (10.1).

Let Φ be a continuous flow in a compact metric space X. Take $A, B \in C(X)^d$ and a continuous function $u\colon X \to \mathbb{R}^+$. We define

$$u_t(x) = \int_0^t u(\varphi_s(x))\,ds$$

for each $t > 0$. Given a continuous function $F\colon \mathbb{R}^d \times \mathbb{R}^d \to \mathbb{R}^d$ and $\alpha \in \mathbb{R}^d$, let

$$L_\alpha = \left\{x \in X : \lim_{t\to\infty} F\left(\frac{A_t(x)}{u_t(x)}, \frac{B_t(x)}{u_t(x)}\right) = \alpha\right\},$$

with $A_t(x)$ and $B_t(x)$ as in (10.7). When $F(X,Y) = X * Y^{-1}$ this is simply the set K_α in (10.1). We also consider the multifractal spectrum $\mathcal{G}_u$ defined by

$$\mathcal{G}_u(\alpha) = \dim_u L_\alpha$$

for each $\alpha \in \mathbb{R}^d$.

We want to establish a relation between the BS-dimension of a set L_α and the BS-dimension of the sets

$$K_{\beta,\gamma} = \left\{x \in X : \lim_{t\to\infty} \left(\frac{A_t(x)}{u_t(x)}, \frac{B_t(x)}{u_t(x)}\right) = (\beta,\gamma)\right\},$$

with $\beta, \gamma \in \mathbb{R}^d$. We write

$$\mathcal{H}_u(\beta, \gamma) = \dim_u K_{\beta,\gamma}.$$

For each $q \in \mathbb{R}^{2d}$, let $S_u(q)$ be the unique real number such that

$$P_\Phi\big(\langle q, (A, B)\rangle - S_u(q)u\big) = 0$$

and let μ_q be the equilibrium measure for $\langle q, (A, B)\rangle - S_u(q)u$ (this measure will be unique in our context).

Applying Theorem 10.1 to the spectrum $\mathcal{H}_u$, we obtain the following result.

Theorem 10.2 *Let Φ be a continuous flow in a compact metric space X such that the map $\mu \mapsto h_\mu(\Phi)$ is upper semicontinuous. If* $\operatorname{span}\{a_1, b_1, \dots, a_d, b_d, u\} \subset D(X)$, *then*

$$\mathcal{H}_u(\partial_q S_u(q)) = S_u(q) - \langle q, \partial_q S_u(q)\rangle$$

and $\mu_q(K_{\partial_q S_u(q)}) = 1$ for every $q \in \mathbb{R}^{2d}$.

Now we describe a general relation between the spectra $\mathcal{G}_u$ and $\mathcal{H}_u$.

Proposition 10.1 ([6]) *Let Φ be a continuous flow in a compact metric space X and let $F\colon \mathbb{R}^d \times \mathbb{R}^d \to \mathbb{R}^d$ be a continuous function. Then*

$$\mathcal{G}_u(\alpha) \ge \sup\big\{\mathcal{H}_u(\beta, \gamma) : (\beta, \gamma) \in F^{-1}(\alpha)\big\}$$

for every α in the image of F.

Proof Take $(\beta, \gamma) \in \mathbb{R}^d \times \mathbb{R}^d$ and $x \in K_{\beta,\gamma}$. By the continuity of F, we have

$$\lim_{t\to\infty} F\left(\frac{A_t(x)}{u_t(x)}, \frac{B_t(x)}{u_t(x)}\right) = F\left(\lim_{t\to\infty} \frac{A_t(x)}{u_t(x)}, \lim_{t\to\infty} \frac{B_t(x)}{u_t(x)}\right) = F(\beta, \gamma).$$

This implies that

$$\bigcup_{(\beta,\gamma)\in F^{-1}(\alpha)} K_{\beta,\gamma} \subseteq L_\alpha.$$

Since $K_{\beta,\gamma} \subset L_\alpha$ for every $(\beta, \gamma) \in F^{-1}(\alpha)$, we have

$$\dim_u K_{\beta,\gamma} \le \dim_u L_\alpha.$$

This yields the desired inequality. □

Barreira, Saussol and Schmeling in [16] made a corresponding study in the case of discrete time.

10.3 Hyperbolic Flows: Analyticity of the Spectrum

In this section we consider the particular case of hyperbolic flows and we establish the analyticity of the spectrum $\mathcal{F}_u$ in (10.2). The proof is based on property 2 of Theorem 10.1 saying that the spectrum is equal to the minimum of a certain function defined implicitly in terms of the topological pressure.

Theorem 10.3 ([6]) *Let Φ be a C^1 flow with a compact locally maximal hyperbolic set Λ such that $\Phi|\Lambda$ is topologically mixing. If the functions $a_i, b_i : \Lambda \to \mathbb{R}$ for $i = 1, \dots, d$ and $u : \Lambda \to \mathbb{R}^+$ are Hölder continuous, then $\mathcal{F}_u$ is analytic in* $\operatorname{int}\mathcal{P}(\mathcal{M})$.

Proof By Proposition 4.4, the map $\mu \mapsto h_\mu(\Phi)$ is upper semicontinuous in $\mathcal{M}$. Thus, by Theorem 10.1, we have

$$\mathcal{F}_u(\alpha) = \min\{T_u(\alpha, q) : q \in \mathbb{R}^d\},$$

where $T_u(\alpha, q)$ is the unique real number satisfying (10.5). Hence,

$$\begin{aligned} 0 &= \partial_q P_\Phi(\langle q, A - \alpha * B\rangle - T_u(\alpha, q)u) \\ &= \partial_q P_\Phi(\langle q, A - \alpha * B\rangle - pu)|_{p=T_u(\alpha,q)} \\ &\quad + \partial_p P_\Phi(\langle q, A - \alpha * B\rangle - pu)|_{p=T_u(\alpha,q)} \partial_q T_u(\alpha, q). \end{aligned}$$

Now take $q(\alpha) \in \mathbb{R}^d$ such that $\mathcal{F}_u(\alpha) = T_u(\alpha, q(\alpha))$. Since T_u is of class C^1 (see the discussion at the end of Sect. 10.1), we have $\partial_q T_u(\alpha, q(\alpha)) = 0$ and thus,

$$\partial_q P_\Phi\big(\langle q, A - \alpha * B\rangle - pu\big) = 0$$

for $q = q(\alpha)$ and $p = T_u(\alpha, q(\alpha))$. Hence, $(\alpha, q, p) = (\alpha, q(\alpha), \mathcal{F}_u(\alpha))$ is a solution of the system

$$\begin{cases} P_\Phi\big(\langle q, A - \alpha * B\rangle - pu\big) = 0, \\ \partial_q P_\Phi\big(\langle q, A - \alpha * B\rangle - pu\big) = 0. \end{cases} \tag{10.8}$$

By Proposition 4.4, the function $t \mapsto P_\Phi(a + tb)$ is analytic. We want to show that

$$\det\left(\frac{\partial\big[P_\Phi\big(\langle q, A - \alpha * B\rangle - pu\big), \partial_q P_\Phi\big(\langle q, A - \alpha * B\rangle - pu\big)\big]}{\partial(q, p)}\right) \neq 0 \tag{10.9}$$

for $(\alpha, q, p) = (\alpha, q(\alpha), \mathcal{F}_u(\alpha))$. The first line of the matrix in (10.9) is

$$\left(\partial_q\big(P_\Phi\big(\langle q, A - \alpha * B\rangle - pu\big)\big), -\int_\Lambda u\, d\mu_\alpha\right),$$

where $\mu_\alpha \in \mathcal{M}$ is the equilibrium measure for $\langle q(\alpha), A - \alpha * B\rangle - \mathcal{F}_u(\alpha)u$. Now we observe that in the last d equations of system (10.8) all values of the first line vanish at $(\alpha, q(\alpha), \mathcal{F}_u(\alpha))$, except for the last one, which is negative. Therefore, the determinant in (10.9) is nonzero provided that

$$\det\left[\partial_q^2 P_\Phi\left(\langle q, A-\alpha * B\rangle - pu\right)\right] \neq 0 \tag{10.10}$$

for $(\alpha, q, p) = (\alpha, q(\alpha), \mathcal{F}_u(\alpha))$.

Lemma 10.1 *The matrix*

$$\partial_q^2 P_\Phi\left(\langle q, A-\alpha * B\rangle - pu\right) \tag{10.11}$$

is positive definite for every $q \in \mathbb{R}^d$, $p \in \mathbb{R}$ and $\alpha \in \operatorname{int} \mathcal{P}(\mathcal{M})$.

Proof of the lemma If the determinant of the matrix in (10.11) is zero, then there exists a vector $v \in \mathbb{R}^d \setminus \{0\}$ such that

$$v^* \partial_q^2 P_\Phi\left(\langle q, A-\alpha * B\rangle - pu\right) v = 0,$$

where v^* is the transpose of v. Then

$$\partial_t^2 P_\Phi\left(\langle q - tv, A-\alpha * B\rangle - pu\right)|_{t=0} = 0$$

and by Proposition 4.4, the function $\langle v, A-\alpha * B\rangle$ is Φ-cohomologous to a constant, say c. Therefore,

$$\int_\Lambda \langle v, A-\alpha * B\rangle \, d\mu = \left\langle \int_\Lambda A \, d\mu - \alpha * \int_\Lambda B \, d\mu \right\rangle = c\mu(\Lambda)$$

for $\mu \in \mathcal{M}$. Since $\alpha \in \mathcal{P}(\mathcal{M})$, we obtain $c = 0$. Hence, the function $\langle v, A-\alpha * B\rangle$ is Φ-cohomologous to 0 and

$$P_\Phi(0) = P_\Phi(t\langle v, A-\alpha * B\rangle) \quad \text{for} \quad t \in \mathbb{R}.$$

Since $\alpha \in \operatorname{int} \mathcal{P}(\mathcal{M})$, there exist $s \neq 0$ and $\mu_s \in \mathcal{M}$ such that $sv + \alpha \in \mathcal{P}(\mathcal{M})$ and

$$\int_\Lambda A \, d\mu_s = \int_\Lambda (sv + \alpha) * B \, d\mu_s.$$

For each $t \in \mathbb{R}$, we obtain

$$\begin{aligned} P_\Phi(0) &= P_\Phi(t\langle sv, A-\alpha * B\rangle) \\ &\geq h_{\mu_s}(\Phi) + t\left\langle sv, (sv + \alpha - \alpha) * \int_\Lambda B \, d\mu_s \right\rangle \\ &\geq ts^2 |v|^2 \inf_{i \in \{1,\ldots,d\}} \inf b_i. \end{aligned}$$

But letting $t \to \infty$, we find that this is impossible, and hence, the matrix in (10.11) has nonzero determinant. Now we show that it is positive definite. By the continuity of the map

$$v \mapsto v^* \partial_q^2 P_\Phi\bigl(\langle q, A - \alpha * B\rangle - pu\bigr)v,$$

if there exist vectors $v = (v_1, \dots, v_d)$ and $w = (w_1, \dots, w_d)$ in $\mathbb{R}^d \setminus \{0\}$ such that

$$v^* \partial_q^2 P_\Phi\bigl(\langle q, A - \alpha * B\rangle - pu\bigr)v < 0$$

and

$$w^* \partial_q^2 P_\Phi\bigl(\langle q, A - \alpha * B\rangle - pu\bigr)w > 0,$$

then one can find $t_1, \dots, t_d \in (0, 1)$ such that

$$x = \bigl(t_1 v_1 + (1 - t_1)w_1, \dots, t_d v_d + (1 - t_d)w_d\bigr) \neq 0$$

and

$$x^* \partial_q^2 P_\Phi\bigl(\langle q, A - \alpha * B\rangle - pu\bigr)x = 0.$$

But it was shown above that this is impossible. Therefore, the matrix in (10.11) is either positive definite or negative definite.

Let e_1 be the first element of the canonical base of $\mathbb{R}^d$. By Proposition 4.4, we have

$$e_1^* \partial_q^2 P_\Phi\bigl(\langle q, A - \alpha * B\rangle - pu\bigr)e_1 = \frac{\partial^2}{\partial q_1^2} P_\Phi\bigl(\langle q, A - \alpha * B\rangle - pu\bigr) \geq 0.$$

This shows that the matrix in (10.11) is positive definite. □

By Lemma 10.1, condition (10.10) holds. Thus, by the Implicit function theorem, system (10.8) defines q and p as analytic functions of α in a neighborhood of $(\alpha, q(\alpha), \mathcal{F}_u(\alpha))$. In particular, the spectrum $\mathcal{F}_u$ is analytic in $\operatorname{int} \mathcal{P}(\mathcal{M})$. □

Chapter 11
Dimension Spectra

In this chapter, for conformal flows with a hyperbolic set, we establish a conditional variational principle for the dimension spectra of Birkhoff averages. The main novelty in comparison to the former chapters is that we consider simultaneously Birkhoff averages into the future and into the past. The main difficulty is that even though the local product structure is bi-Lipschitz, the level sets of the Birkhoff averages are not compact. Our proof is based on the use of Markov systems and is inspired by earlier arguments in the case of discrete time.

11.1 Future and Past

In this section we consider Birkhoff averages both into the future and into the past, and we compute the Hausdorff dimension of the corresponding level sets on locally maximal hyperbolic sets for a conformal flow.

Let $\Phi = (\varphi_t)_{t\in\mathbb{R}}$ be a C^1 flow with a locally maximal hyperbolic set Λ such that $\Phi|\Lambda$ is conformal. We denote by $C^\gamma(\Lambda)$ the space of Hölder continuous functions in Λ with Hölder exponent $\gamma \in (0,1)$. Given $d \in \mathbb{N}$, let $F = C^\gamma(\Lambda)^d \times C^\gamma(\Lambda)^d$. Moreover, given functions $(A^\pm, B^\pm) \in F$, we write

$$A^+ = (a_1^+, \dots, a_d^+), \quad B^+ = (b_1^+, \dots, b_d^+) \tag{11.1}$$

and

$$A^- = (a_1^-, \dots, a_d^-), \quad B^- = (b_1^-, \dots, b_d^-), \tag{11.2}$$

and we assume that all components of A^- and B^- are positive functions. For each $\alpha = (\alpha_1, \dots, \alpha_d)$ and $\beta = (\beta_1, \dots, \beta_d)$ in $\mathbb{R}^d$, let

$$K_\alpha^+ = \bigcap_{i=1}^d \left\{ x \in \Lambda : \lim_{t\to+\infty} \frac{\int_0^t a_i^+(\varphi_s(x))\,ds}{\int_0^t b_i^+(\varphi_s(x))\,ds} = \alpha_i \right\}$$

L. Barreira, *Dimension Theory of Hyperbolic Flows*,
Springer Monographs in Mathematics, DOI 10.1007/978-3-319-00548-5_11,

and

$$K_\beta^- = \bigcap_{i=1}^{d} \left\{ x \in \Lambda : \lim_{t\to-\infty} \frac{\int_0^t a_i^-(\varphi_s(x))\,ds}{\int_0^t b_i^-(\varphi_s(x))\,ds} = \beta_i \right\}.$$

The following result expresses the dimensions of the level sets K_α^+ and K_β^- in terms of the topological pressure.

Theorem 11.1 ([7]) *Let Φ be a $C^{1+\delta}$ flow with a locally maximal hyperbolic set Λ such that $\Phi|\Lambda$ is conformal and topologically mixing and let $(A^\pm, B^\pm) \in F$. For each $\alpha, \beta \in \mathbb{R}^d$, $x^+ \in K_\alpha^+$ and $x^- \in K_\beta^-$, we have*

$$\begin{aligned} \dim_H K_\alpha^+ &= \dim_H(K_\alpha^+ \cap V^u(x^+)) + t_s + 1 \\ &= \dim_{\zeta_u} K_\alpha^+ + t_s + 1 \end{aligned} \tag{11.3}$$

and

$$\begin{aligned} \dim_H K_\beta^- &= \dim_H(K_\beta^- \cap V^s(x^-)) + t_u + 1 \\ &= \dim_{-\zeta_s} K_\beta^- + t_u + 1, \end{aligned} \tag{11.4}$$

with t_s and t_u as in (5.6).

Proof By (3.1) and the uniform continuity of $a_i^\pm$ and $b_i^\pm$ in Λ, we have

$$\Lambda \cap V^s(x) \subset K_\alpha^+ \quad \text{for} \quad x \in K_\alpha^+,$$

and thus,

$$\Lambda \cap \bigcup_{t\in\mathbb{R}} \varphi_t(V^s(x)) \subset K_\alpha^+$$

for $x \in K_\alpha^+$, since the set K_α^+ is Φ-invariant.

On the other hand, since Φ is conformal on Λ, it follows from results of Hasselblatt in [53] that the distributions $x \mapsto E^s(x) \oplus E^0(x)$ and $x \mapsto E^u(x) \oplus E^0(x)$ are Lipschitz. Therefore, on a sufficiently small open neighborhood of a point $x \in K_\alpha^+$ there exists a Lipschitz map with Lipschitz inverse from the set K_α^+ to the product

$$\bigcup_{t\in I} \varphi_t(V^s(x)) \times V^u(x),$$

where I is some open interval containing zero. Therefore,

$$\dim_H K_\alpha^+ = \dim_H\left((K_\alpha^+ \cap V^u(x)) \times \left(\Lambda \cap \bigcup_{t\in I} \varphi_t(V^s(x)) \right) \right). \tag{11.5}$$

On the other hand, by Theorem 5.1, we have

$$\dim_H\left(\Lambda\cap\bigcup_{t\in I}\varphi_t(V^s(x))\right)=\overline{\dim}_B\left(\Lambda\cap\bigcup_{t\in I}\varphi_t(V^s(x))\right)=t_s+1. \tag{11.6}$$

Since

$$\dim_H E+\dim_H F\le\dim_H(E\times F)\le\dim_H E+\overline{\dim}_B F$$

for any sets $E,F\subset\mathbb{R}^m$ (see for example [41]), it follows from (11.5) and (11.6) that

$$\dim_H K_\alpha^+=\dim_H(K_\alpha^+\cap V^u(x))+t_s+1.$$

For the second equality in (11.3), we note that

$$\int_0^t\zeta_u(\varphi_s(x))\,ds=\log\|d_x\varphi_t|E^u(x)\|.$$

Since the distribution $x\mapsto E^u(x)\oplus E^0(x)$ is Lipschitz and Φ is of class $C^{1+\delta}$, the function ζ_u is Hölder continuous and for each $\varepsilon>0$ there exist constants $c_1,c_2>0$ such that

$$c_1\exp(-\alpha\zeta_u(x,t,\varepsilon))\le\left[\operatorname{diam}\big(B(x,t,\varepsilon)\cap V^u(x)\big)\right]^\alpha\le c_2\exp(-\alpha\zeta_u(x,t,\varepsilon))$$

for every $x\in\Lambda$ and $t>0$. Hence, it follows from the definition of Hausdorff dimension that

$$\dim_H(Z\cap V^u(x))=\dim_{\zeta_u}Z$$

for every set $Z\subset\Lambda$. The second equality in (11.3) is obtained by taking $Z=K_\alpha^+$.

The arguments for the set K_β^- and (11.4) are entirely analogous. □

11.2 Conditional Variational Principle

In this section we establish a conditional variational principle for the dimension spectrum obtained from the level sets $K_\alpha^+\cap K_\beta^-$.

Definition 11.1 The *dimension spectrum* $\mathcal{D}\colon\mathbb{R}^d\times\mathbb{R}^d\to\mathbb{R}$ associated to the functions in (11.1) and (11.2) is defined by

$$\mathcal{D}(\alpha,\beta)=\dim_H(K_\alpha^+\cap K_\beta^-).$$

The following result is a conditional variational principle for the spectrum $\mathcal{D}$.

Theorem 11.2 ([7]) *Let Φ be a $C^{1+\delta}$ flow with a locally maximal hyperbolic set Λ such that $\Phi|\Lambda$ is conformal and topologically mixing and let $(A^\pm,B^\pm)\in F$. Then the following properties hold*:

1. *if*

$$\alpha \in \operatorname{int} \mathcal{P}^+(\mathcal{M}) \quad \textit{and} \quad \beta \in \operatorname{int} \mathcal{P}^-(\mathcal{M}), \tag{11.7}$$

then

$$\begin{aligned}\mathcal{D}(\alpha, \beta) &= \dim_H K_\alpha^+ + \dim_H K_\beta^- - \dim_H \Lambda \\ &= \max\left\{\frac{h_\mu(\Phi)}{\int_\Lambda \zeta_u \, d\mu} : \mu \in \mathcal{M} \textit{ and } \mathcal{P}^+(\mu) = \alpha\right\} \\ &\quad + \max\left\{\frac{h_\mu(\Phi)}{-\int_\Lambda \zeta_s \, d\mu} : \mu \in \mathcal{M} \textit{ and } \mathcal{P}^-(\mu) = \beta\right\} + 1;\end{aligned}$$

2. *the function* $\mathcal{D}$ *is analytic in* $\operatorname{int} \mathcal{P}^+(\mathcal{M}) \times \operatorname{int} \mathcal{P}^-(\mathcal{M})$.

Proof The proof is based on arguments of Barreira and Valls in [18], using also results of Barreira and Saussol in [12]. We separate the argument into several steps.

Step 1. Construction of auxiliary measures

Consider a Markov system $R_1, \ldots, R_k$ for Φ on Λ and the associated symbolic dynamics (see Sect. 3.3). The following statement is a consequence of a construction described by Bowen in [28].

Lemma 11.1 *For* $i, j = 1, \ldots, d$ *there exist Hölder continuous functions*

$$a_i^u, b_i^u, d^u : \Sigma_A^+ \to \mathbb{R} \quad \textit{and} \quad a_j^s, b_j^s, d^s : \Sigma_A^- \to \mathbb{R},$$

and continuous functions $g_i^+, h_i^+, g_j^-, h_j^-, \rho^\pm : \Sigma_A \to \mathbb{R}$ *such that*

$$I_{a_i^+} \circ \pi = a_i^u \circ \pi_+ + g_i^+ - g_i^+ \circ \sigma,$$

$$I_{b_i^+} \circ \pi = b_i^u \circ \pi_+ + h_i^+ - h_i^+ \circ \sigma,$$

$$I_{\zeta_u} \circ \pi = d^u \circ \pi_+ + \rho^+ - \rho^+ \circ \sigma$$

and

$$I_{a_j^-} \circ \pi = a_j^s \circ \pi_- + g_j^- - g_j^- \circ \sigma^{-1},$$

$$I_{b_j^-} \circ \pi = b_j^s \circ \pi_- + h_j^- - h_j^- \circ \sigma^{-1},$$

$$I_{-\zeta_s} \circ \pi = d^s \circ \pi_- + \rho^- - \rho^- \circ \sigma^{-1}.$$

We write

$$A^u=\left(a_1^u,\dots,a_d^u\right),\quad B^u=\left(b_1^u,\dots,b_d^u\right)$$

and

$$A^s=\left(a_1^s,\dots,a_d^s\right),\quad B^s=\left(b_1^s,\dots,b_d^s\right).$$

Given $q^\pm\in\mathbb{R}^d$, we define Hölder continuous functions $U\colon\Sigma_A^+\to\mathbb{R}$ and $S\colon\Sigma_A^-\to\mathbb{R}$ by

$$\begin{aligned}U&=\langle q^+,A^u-\alpha*B^u\rangle-d^+d^u,\\ S&=\langle q^-,A^s-\beta*B^s\rangle-d^-d^s,\end{aligned}\tag{11.8}$$

where

$$d^+=\dim_H K_\alpha^+-t_s-1\quad\text{and}\quad d^-=\dim_H K_\beta^--t_u-1.\tag{11.9}$$

Now let μ^u be the equilibrium measure for U in Σ_A^+ (with respect to σ_+) and let μ^s be the equilibrium measure for S in Σ_A^- (with respect to σ_-). The following result is a simple consequence of Theorem 10.1.

Lemma 11.2 *For each α and β as in* (11.7), *there exist $q^\pm\in\mathbb{R}^d$ such that*

$$P_{\sigma_+}(U)=P_{\sigma_-}(S)=0,$$

$$\int_{\Sigma_A^+}A^u\,d\mu^u=\alpha*\int_{\Sigma_A^+}B^u\,d\mu^u$$

and

$$\int_{\Sigma_A^-}A^u\,d\mu^s=\beta*\int_{\Sigma_A^-}B^s\,d\mu^s.$$

Given $x\in Z=\bigcup_{i=1}^k R_i$, let $R(x)$ be a rectangle of the Markov system that contains x. We define measures ν^u and ν^s on $R(x)$ by

$$\nu^u=\mu^u\circ\pi_+\circ\pi^{-1}\quad\text{and}\quad\nu^s=\mu^s\circ\pi_-\circ\pi^{-1},$$

taking the vectors $q^\pm$ given by Lemma 11.2. Finally, we define a measure ν on $R(x)$ by $\nu=\nu^u\times\nu^s$. Since μ^u and μ^s are Gibbs measures (see (7.8)), we have

$$\nu(R(x))=\mu^u\left(C_{i_0}^+\right)\mu^s\left(C_{i_0}^-\right)>0,$$

with $C_{i_0}^+$ and $C_{i_0}^-$ as in (3.15) and (3.16).

Step 2. Lower pointwise dimension

Here and in the following steps we establish several properties of the measure ν.

Lemma 11.3 *For ν-almost every $x \in Z$, we have*

$$\liminf_{r\to 0} \frac{\log \nu(B(x,r))}{\log r} \geq \dim_H K_\alpha^+ + \dim_H K_\beta^- - \dim_H \Lambda - 1.$$

Proof of the lemma We follow arguments in the proof of Lemma 4 in [18]. By the variational principle for the topological pressure applied to the functions U and S in (11.8) together with Lemma 11.2, we obtain

$$\frac{h_{\mu^u}(\sigma_+)}{\int_{\Sigma_A^+} d^u \, d\mu^u} = d^+ \quad \text{and} \quad \frac{h_{\mu^s}(\sigma_-)}{\int_{\Sigma_A^-} d^s \, d\mu^s} = d^-.$$

By the Shannon–McMillan–Breiman theorem and Birkhoff's ergodic theorem, given $\varepsilon > 0$, for μ^s-almost every $\omega^+ \in C_{i_0}^+$ and μ^u-almost every $\omega^- \in C_{i_0}^-$ there exists an $s(\omega) \in \mathbb{N}$, with $\omega^+ = \pi_+(\omega)$ and $\omega^- = \pi_-(\omega)$, such that

$$d^+ - \varepsilon < -\frac{\log \mu^u(C_{i_0\cdots i_n}^+)}{\sum_{k=0}^n d^u(\sigma_+^k(\omega^+))} < d^+ + \varepsilon$$

and

$$d^- - \varepsilon < -\frac{\log \mu^s(C_{i_{-m}\cdots i_0}^-)}{\sum_{k=0}^m d^s(\sigma_-^k(\omega^-))} < d^- + \varepsilon$$

for $n, m > s(\omega)$. For any sufficiently small $r > 0$, let $n = n(\omega, r)$ and $m = m(\omega, r)$ be the unique positive integers such that

$$-\sum_{k=0}^{n} d^u(\sigma_+^k(\omega^+)) > \log r, \quad -\sum_{k=0}^{n+1} d^u(\sigma_+^k(\omega^+)) \leq \log r \tag{11.10}$$

and

$$-\sum_{k=0}^{m} d^s(\sigma_-^k(\omega^-)) > \log r, \quad -\sum_{k=0}^{m+1} d^s(\sigma_-^k(\omega^-)) \leq \log r. \tag{11.11}$$

On the other hand, as in the proof of Theorem 8.3 (see (8.21)), there exists a $\rho > 1$ (independent of $x = \pi(\omega)$ and r) such that

$$B(y, r/\rho) \cap Z \subset \pi(C_{i_{-m}\cdots i_n}) \subset B(x, \rho r) \cap Z \tag{11.12}$$

for some point $y \in \pi(C_{i_{-m}\cdots i_n})$, where $\omega = (\cdots i_{-1} i_0 i_1 \cdots)$. Now we recall a result of Barreira and Saussol in [13] (see also Lemma 15.2.2 in [3]).

Lemma 11.4 *Given a probability measure ν on a set $Z \subset \mathbb{R}^m$, there exists a constant $\eta > 1$ such that for ν-almost every $y \in Z$ and every $\varepsilon > 0$ there exists a $c = c(y, \varepsilon)$ such that*

$$\nu(B(y, \eta r)) \leq \nu(B(y,r)) r^{-\varepsilon} \quad \textit{for} \quad r < c.$$

Without loss of generality, we take $\eta = 2\rho$. By (11.12) and Lemma 11.4, we obtain

$$
\begin{aligned}
\nu(B(x,r)) &\le \nu\left(B(y, 2\rho\frac{r}{\rho})\right) \le \nu(B(y, r/\rho))\left(\frac{r}{\rho}\right)^{-\varepsilon} \\
&\le \nu(\pi(C_{i_{-m}\cdots i_n}))\left(\frac{r}{\rho}\right)^{-\varepsilon} = \mu^u(C^+_{i_0\cdots i_n})\mu^s(C^-_{i_{-m}\cdots i_0})\left(\frac{r}{\rho}\right)^{-\varepsilon} \\
&\le \exp\left[(-d^+ + \varepsilon)\sum_{k=0}^{n} d^u(\sigma_+^k(\omega^+))\right] \\
&\quad \times \exp\left[(-d^- + \varepsilon)\sum_{k=0}^{m} d^s(\sigma_-^k(\omega^-))\right]\left(\frac{r}{\rho}\right)^{-\varepsilon} \\
&\le \exp[(\log r + \|d^u\|_\infty)(d^+ - \varepsilon)]\exp[(\log r + \|d^s\|_\infty)(d^- - \varepsilon)]\left(\frac{r}{\rho}\right)^{-\varepsilon}
\end{aligned}
$$

for $r < c$, and hence,

$$\liminf_{r\to 0} \frac{\log \nu(B(x,r))}{\log r} \ge d^+ + d^- - 2\varepsilon \tag{11.13}$$

for ν-almost every $x \in Z$.

On the other hand, by Theorem 5.2, we have

$$\dim_H \Lambda = t_s + t_u + 1. \tag{11.14}$$

Thus, by (11.9) and (11.13), we obtain

$$\liminf_{r\to 0} \frac{\log \nu(B(x,r))}{\log r} \ge \dim_H K_\alpha^+ + \dim_H K_\beta^- - \dim_H \Lambda - 1 - 2\varepsilon$$

and the desired result follows from the arbitrariness of ε. □

Step 3. Upper pointwise dimension

Now we obtain an upper bound for the upper pointwise dimension.

Lemma 11.5 *For each $x \in K_\alpha^+ \cap K_\beta^- \cap Z$, we have*

$$\limsup_{r\to 0} \frac{\log \nu(B(x,r))}{\log r} \le \dim_H K_\alpha^+ + \dim_H K_\beta^- - \dim_H \Lambda - 1.$$

Proof of the lemma We follow arguments in the proofs of Lemmas 5 and 6 in [18]. Take $x \in K_\alpha^+ \cap K_\beta^- \cap Z$ and $\omega \in \Sigma_A$ such that $\pi(\omega) = x$, and let $\omega^\pm = \pi^\pm(\omega)$.

It follows from Lemma 11.1 that

$$\begin{aligned} I_{a_i^+}(T^k(\pi(\omega))) &= I_{a_i^+}(\pi(\sigma_+^k(\omega))) \\ &= a_i^u(\pi_+(\sigma^k(\omega))) + g_i^+(\sigma^k(\omega)) - g_i^+(\sigma^{k+1}(\omega)) \\ &= a_i^u(\sigma_+^k(\omega^+)) + g_i^+(\sigma^k(\omega)) - g_i^+(\sigma^{k+1}(\omega)), \end{aligned}$$

with analogous identities for the functions $I_{b_i^+}$, $I_{a_j^-}$ and $I_{b_j^-}$. Therefore,

$$\frac{\sum_{k=0}^{n-1} I_{a_i^+}(T^k(x))}{\sum_{k=0}^{n-1} I_{b_i^+}(T^k(x))} = \frac{\sum_{k=0}^{n-1} a_i^u(\sigma_+^k(\omega^+)) + g_i^+(\omega) - g_i^+(\sigma^n(\omega))}{\sum_{k=0}^{n-1} b_i^u(\sigma_+^k(\omega^+)) + h_i^+(\omega) - h_i^+(\sigma^n(\omega))}$$

and

$$\frac{\sum_{k=0}^{n-1} I_{a_j^-}(T^{-k}(x))}{\sum_{k=0}^{n-1} I_{b_j^-}(T^{-k}(x))} = \frac{\sum_{k=0}^{n-1} a_j^s(\sigma_-^k(\omega^-)) + g_j^-(\omega) - g_j^-(\sigma^{-n}(\omega))}{\sum_{k=0}^{n-1} b_j^s(\sigma_-^k(\omega^-)) + h_j^-(\omega) - h_j^-(\sigma^{-n}(\omega))}.$$

On the other hand,

$$\sum_{k=0}^{n-1} b_i^u(\sigma_+^k(\omega^+)) \geq n \inf b_i^+ \inf \tau - 2\|h_i^+\|_\infty$$

and

$$\sum_{k=0}^{n-1} b_j^s(\sigma_-^k(\omega^-)) \geq n \inf b_j^- \inf \tau - 2\|h_j^-\|_\infty.$$

Since $b_i^+, b_j^- > 0$ and $\inf \tau > 0$, this ensures that the limits

$$\lim_{n\to\infty} \frac{\sum_{k=0}^{n-1} I_{a_i^+}(T^k(x))}{\sum_{k=0}^{n-1} I_{b_i^+}(T^k(x))}, \quad \lim_{n\to\infty} \frac{\sum_{k=0}^{n-1} I_{a_j^-}(T^{-k}(x))}{\sum_{k=0}^{n-1} I_{b_j^-}(T^{-k}(x))}$$

exist respectively if and only if the limits

$$\lim_{n\to\infty} \frac{\sum_{k=0}^{n-1} a_i^u(\sigma_+^k(\omega^+))}{\sum_{k=0}^{n-1} b_i^u(\sigma_+^k(\omega^+))}, \quad \lim_{n\to\infty} \frac{\sum_{k=0}^{n-1} a_j^s(\sigma_-^k(\omega^-))}{\sum_{k=0}^{n-1} b_j^s(\sigma_-^k(\omega^-))}$$

exist, in which case they (respectively) coincide.

By Theorem 2.3, if $x \in K_\alpha^+ \cap K_\beta^- \cap Z$ and $\omega \in \Sigma_A$ are such that $\pi(\omega) = x$, then given $\varepsilon > 0$, there exists an $r(\omega) \in \mathbb{N}$ such that

$$\left\| \left\langle q^+, \sum_{k=0}^{n} (A^u - \alpha * B^u)(\sigma_+^k(\omega^+)) \right\rangle \right\| < \varepsilon n \|\langle q^+, B^u \rangle\|_\infty$$

and

$$\left\| \left\langle q^-, \sum_{k=0}^{n} (A^s - \beta * B^s)(\sigma_-^k(\omega^-)) \right\rangle \right\| < \varepsilon n \|\langle q^-, B^s\rangle\|_\infty$$

for $n > r(\omega)$. By Lemma 11.2, we have $P_{\sigma_+}(U) = 0$ and since μ^u is a Gibbs measure (see (7.8)), there exists a $D > 0$ such that

$$D^{-1} < \frac{\mu^u(C^+_{i_0\cdots i_n})}{\exp \sum_{k=0}^{n} U(\sigma_+^k(\omega^+))} < D$$

for $i_0 = 1, \dots, p$ and $n \in \mathbb{N}$. Therefore,

$$\mu^u(C^+_{i_0\cdots i_n}) > D^{-1} \exp\left[-d^+ \sum_{k=0}^{n} d^u(\sigma_+^k(\omega^+)) - \varepsilon n \|\langle q^+, B^u\rangle\|_\infty \right]. \tag{11.15}$$

Similarly, for every $i_0 = 1, \dots, p$ and $n \in \mathbb{N}$ we have

$$\mu^s(C^-_{i_{-m}\cdots i_0}) > D^{-1} \exp\left[-d^- \sum_{k=0}^{m} d^s(\sigma_-^k(\omega^-)) - \varepsilon m \|\langle q^-, B^s\rangle\|_\infty \right]. \tag{11.16}$$

Since $\inf_{x\in\Lambda} \tau > 0$ (see (3.7)), it follows from the hyperbolicity of Φ on Λ that there exists an $r > 0$ such that $n(\omega, r) > r(\omega)$ and $m(\omega, r) > r(\omega)$ (see (11.10) and (11.11)). Moreover, by (11.12), there exists a $\rho > 0$ (independent of $x = \pi(\omega)$ and r) such that

$$B(x, \rho r) \cap Z \supset \pi(C_{i_{-m}\cdots i_n}),$$

where $n = n(\omega, r)$ and $m = m(\omega, r)$. Combining (11.15) and (11.16) with (11.10) and (11.11), we obtain

$$\begin{aligned} \nu(B(x, \rho r)) &\geq \nu(\pi(C_{i_{-m}\cdots i_n})) \\ &= \mu^u(C^+_{i_0\cdots i_n}) \mu^s(C^-_{i_{-m}\cdots i_0}) \\ &\geq D^{-2} r^{d^+ + d^-} \exp\bigl(-\varepsilon n \|\langle q^+, B^u\rangle\|_\infty - \varepsilon m \|\langle q^-, B^s\rangle\|_\infty\bigr) \end{aligned}$$

for any sufficiently small $r > 0$. On the other hand, it follows from (11.10) and (11.11) that

$$-n \inf d^u > \log r \quad \text{and} \quad -m \inf d^s > \log r.$$

Therefore, for each $x \in K^+_\alpha \cap K^-_\beta \cap Z$ we have

$$\limsup_{r\to\infty} \frac{\log \nu(B(x, r))}{\log r} \leq d^+ + d^- + \varepsilon \left(\frac{\|\langle q^+, B^u\rangle\|_\infty}{\inf d^u} + \frac{\|\langle q^-, B^s\rangle\|_\infty}{\inf d^s} \right).$$

Since ε can be made arbitrarily small, we obtain

$$\limsup_{r\to\infty} \frac{\log \nu(B(x,r))}{\log r} \le d^+ + d^-.$$

Together with (11.9) and (11.14) this yields the desired result. □

Step 4. Conclusion

Combining Lemmas 11.3 and 11.5 yields the following result.

Lemma 11.6 *For each α and β as in* (11.7), *there exists a probability measure ν in Z such that $\nu(K_\alpha^+ \cap K_\beta^-) = 1$,*

$$\lim_{r\to\infty} \frac{\log \nu(B(x,r))}{\log r} = \dim_H K_\alpha^+ + \dim_H K_\beta^- - \dim_H \Lambda - 1 \tag{11.17}$$

for ν-almost every $x \in Z$, and

$$\limsup_{r\to\infty} \frac{\log \nu(B(x,r))}{\log r} \le \dim_H K_\alpha^+ + \dim_H K_\beta^- - \dim_H \Lambda - 1 \tag{11.18}$$

for every $x \in K_\alpha^+ \cap K_\beta^- \cap Z$.

We proceed with the proof of the theorem. It follows from (11.17) (see for example [3, Theorem 2.1.5]) that

$$\dim_H \nu = \dim_H K_\alpha^+ + \dim_H K_\beta^- - \dim_H \Lambda - 1,$$

where

$$\dim_H \nu = \inf\{\dim_H Z : \nu(Z) = 1\}.$$

Since $\nu(K_\alpha^+ \cap K_\beta^-) = 1$, we obtain

$$\dim_H (K_\alpha^+ \cap K_\beta^- \cap Z) \ge \dim_H K_\alpha^+ + \dim_H K_\beta^- - \dim_H \Lambda - 1.$$

On the other hand, it follows from (11.18) (see for example [3, Theorem 2.1.5]) that

$$\dim_H (K_\alpha^+ \cap K_\beta^- \cap Z) \le \dim_H K_\alpha^+ + \dim_H K_\beta^- - \dim_H \Lambda - 1,$$

and thus,

$$\dim_H (K_\alpha^+ \cap K_\beta^- \cap Z) = \dim_H K_\alpha^+ + \dim_H K_\beta^- - \dim_H \Lambda - 1.$$

Since $K_\alpha^+ \cap K_\beta^-$ is locally diffeomorphic to a product of $K_\alpha^+ \cap K_\beta^- \cap Z$ and an interval, we obtain

$$\mathcal{D}(\alpha,\beta) = \dim_H K_\alpha^+ + \dim_H K_\beta^- - \dim_H \Lambda.$$

By Theorems 10.1 and 11.1 together with (11.14), we conclude that

$$\begin{aligned}
\mathcal{D}(\alpha, \beta) &= \dim_H(K_\alpha^+ \cap V^u(x)) + \dim_H(K_\beta^- \cap V^s(x)) + 1 \\
&= \dim_{\zeta_u} K_\alpha^+ + \dim_{-\zeta_s} K_\beta^- + 1 \\
&= \max\left\{ \frac{h_\mu(\Phi)}{\int_\Lambda \zeta_u \, d\mu} : \mu \in \mathcal{M} \text{ and } \mathcal{P}^+(\mu) = \alpha \right\} \\
&\quad + \max\left\{ \frac{h_\mu(\Phi)}{-\int_\Lambda \zeta_s \, d\mu} : \mu \in \mathcal{M} \text{ and } \mathcal{P}^-(\mu) = \beta \right\} + 1.
\end{aligned}$$

The second statement is now a simple consequence of Theorem 10.1. □

References

1. L. Abramov, *On the entropy of a flow*, Dokl. Akad. Nauk SSSR **128** (1959), 873–875.
2. L. Barreira, *A non-additive thermodynamic formalism and applications to dimension theory of hyperbolic dynamical systems*, Ergod. Theory Dyn. Syst. **16** (1996), 871–927.
3. L. Barreira, *Dimension and Recurrence in Hyperbolic Dynamics*, Progress in Mathematics 272, Birkhäuser, Basel, 2008.
4. L. Barreira, *Thermodynamic Formalism and Applications to Dimension Theory*, Progress in Mathematics 294, Birkhäuser, Basel, 2011.
5. L. Barreira, *Ergodic Theory, Hyperbolic Dynamics and Dimension Theory*, Universitext, Springer, Berlin, 2012.
6. L. Barreira and P. Doutor, *Birkhoff averages for hyperbolic flows: variational principles and applications*, J. Stat. Phys. **115** (2004), 1567–1603.
7. L. Barreira and P. Doutor, *Dimension spectra of hyperbolic flows*, J. Stat. Phys. **136** (2009), 505–525.
8. L. Barreira and K. Gelfert, *Multifractal analysis for Lyapunov exponents on nonconformal repellers*, Commun. Math. Phys. **267** (2006), 393–418.
9. L. Barreira and G. Iommi, *Suspension flows over countable Markov shifts*, J. Stat. Phys. **124** (2006), 207–230.
10. L. Barreira and Ya. Pesin, *Nonuniform Hyperbolicity: Dynamics of Systems with Nonzero Lyapunov Exponents*, Encyclopedia of Mathematics and Its Applications 115, Cambridge University Press, Cambridge, 2007.
11. L. Barreira, Ya. Pesin and J. Schmeling, *Dimension and product structure of hyperbolic measures*, Ann. Math. (2) **149** (1999), 755–783.
12. L. Barreira and B. Saussol, *Multifractal analysis of hyperbolic flows*, Commun. Math. Phys. **214** (2000), 339–371.
13. L. Barreira and B. Saussol, *Hausdorff dimension of measures via Poincaré recurrence*, Commun. Math. Phys. **219** (2001), 443–463.
14. L. Barreira and B. Saussol, *Variational principles and mixed multifractal spectra*, Trans. Am. Math. Soc. **353** (2001), 3919–3944.
15. L. Barreira and B. Saussol, *Variational principles for hyperbolic flows*, Fields Inst. Commun. **31** (2002), 43–63.
16. L. Barreira, B. Saussol and J. Schmeling, *Higher-dimensional multifractal analysis*, J. Math. Pures Appl. **81** (2002), 67–91.
17. L. Barreira and J. Schmeling, *Sets of "non-typical" points have full topological entropy and full Hausdorff dimension*, Isr. J. Math. **116** (2000), 29–70.
18. L. Barreira and C. Valls, *Multifractal structure of two-dimensional horseshoes*, Commun. Math. Phys. **266** (2006), 455–470.

L. Barreira, *Dimension Theory of Hyperbolic Flows*,
Springer Monographs in Mathematics, DOI 10.1007/978-3-319-00548-5,

19. L. Barreira and C. Wolf, *Measures of maximal dimension for hyperbolic diffeomorphisms*, Commun. Math. Phys. **239** (2003), 93–113.
20. L. Barreira and C. Wolf, *Pointwise dimension and ergodic decompositions*, Ergod. Theory Dyn. Syst. **26** (2006), 653–671.
21. L. Barreira and C. Wolf, *Dimension and ergodic decompositions for hyperbolic flows*, Discrete Contin. Dyn. Syst. **17** (2007), 201–212.
22. T. Bedford, *Crinkly curves, Markov partitions and box dimension of self-similar sets*, Ph.D. Thesis, University of Warwick, 1984.
23. T. Bedford, *The box dimension of self-affine graphs and repellers*, Nonlinearity **2** (1989), 53–71.
24. T. Bedford and M. Urbański, *The box and Hausdorff dimension of self-affine sets*, Ergod. Theory Dyn. Syst. **10** (1990), 627–644.
25. C. Bonatti, L. Díaz and M. Viana, *Discontinuity of the Hausdorff dimension of hyperbolic sets*, C. R. Acad. Sci. Paris Sér. I Math. **320** (1995), 713–718.
26. H. Bothe, *The Hausdorff dimension of certain solenoids*, Ergod. Theory Dyn. Syst. **15** (1995), 449–474.
27. R. Bowen, *Symbolic dynamics for hyperbolic flows*, Am. J. Math. **95** (1973), 429–460.
28. R. Bowen, *Equilibrium States and the Ergodic Theory of Anosov Diffeomorphism*, Lect. Notes in Math. 470, Springer, Berlin, 1975.
29. R. Bowen, *Hausdorff dimension of quasi-circles*, Inst. Hautes Études Sci. Publ. Math. **50** (1979), 259–273.
30. R. Bowen and D. Ruelle, *The ergodic theory of axiom A flows*, Invent. Math. **29** (1975), 181–202.
31. R. Bowen and P. Walters, *Expansive one-parameter flows*, J. Differ. Equ. **12** (1972), 180–193.
32. M. Brin and A. Katok, *On local entropy*, in Geometric Dynamics (Rio de Janeiro, 1981), edited by J. Palis, Lect. Notes in Math. 1007, Springer, Berlin, 1983, pp. 30–38.
33. P. Collet, J. Lebowitz and A. Porzio, *The dimension spectrum of some dynamical systems*, J. Stat. Phys. **47** (1987), 609–644.
34. L. Díaz and M. Viana, *Discontinuity of Hausdorff dimension and limit capacity on arcs of diffeomorphisms*, Ergod. Theory Dyn. Syst. **9** (1989), 403–425.
35. A. Douady and J. Oesterlé, *Dimension de Hausdorff des attracteurs*, C. R. Acad. Sci. Paris **290** (1980), 1135–1138.
36. M. Dysman, *Fractal dimension for repellers of maps with holes*, J. Stat. Phys. **120** (2005), 479–509.
37. K. Falconer, *The Hausdorff dimension of self-affine fractals*, Math. Proc. Camb. Philos. Soc. **103** (1988), 339–350.
38. K. Falconer, *A subadditive thermodynamic formalism for mixing repellers*, J. Phys. A, Math. Gen. **21** (1988), 1737–1742.
39. K. Falconer, *Dimensions and measures of quasi self-similar sets*, Proc. Am. Math. Soc. **106** (1989), 543–554.
40. K. Falconer, *Bounded distortion and dimension for non-conformal repellers*, Math. Proc. Camb. Philos. Soc. **115** (1994), 315–334.
41. K. Falconer, *Fractal Geometry. Mathematical Foundations and Applications*, Wiley, New York, 2003.
42. H. Federer, *Geometric Measure Theory*, Grundlehren der mathematischen Wissenschaften 153, Springer, Berlin, 1969.
43. A. Fan, Y. Jiang and J. Wu, *Asymptotic Hausdorff dimensions of Cantor sets associated with an asymptotically non-hyperbolic family*, Ergod. Theory Dyn. Syst. **25** (2005), 1799–1808.
44. D. Feng, *Lyapunov exponents for products of matrices and multifractal analysis. I. Positive matrices*, Isr. J. Math. **138** (2003), 353–376.
45. D. Feng, *The variational principle for products of non-negative matrices*, Nonlinearity **17** (2004), 447–457.

46. D. Feng and K. Lau, *The pressure function for products of non-negative matrices*, Math. Res. Lett. **9** (2002), 363–378.
47. D. Gatzouras and S. Lalley, *Hausdorff and box dimensions of certain self-affine fractals*, Indiana Univ. Math. J. **41** (1992), 533–568.
48. D. Gatzouras and Y. Peres, *Invariant measures of full dimension for some expanding maps*, Ergod. Theory Dyn. Syst. **17** (1997), 147–167.
49. K. Gelfert, *Dimension estimates beyond conformal and hyperbolic dynamics*, Dyn. Syst. **20** (2005), 267–280.
50. B. Gurevič, *Topological entropy of a countable Markov chain*, Sov. Math. Dokl. **10** (1969), 911–915.
51. T. Halsey, M. Jensen, L. Kadanoff, I. Procaccia and B. Shraiman, *Fractal measures and their singularities: the characterization of strange sets*, Phys. Rev. A **34** (1986), 1141–1151; errata in **34** (1986), 1601.
52. P. Hanus, R. Mauldin and M. Urbański, *Thermodynamic formalism and multifractal analysis of conformal infinite iterated function systems*, Acta Math. Hung. **96** (2002), 27–98.
53. B. Hasselblatt, *Regularity of the Anosov splitting and of horospheric foliations*, Ergod. Theory Dyn. Syst. **14** (1994), 645–666.
54. B. Hasselblatt and J. Schmeling, *Dimension product structure of hyperbolic sets*, Electron. Res. Announc. Am. Math. Soc. **10** (2004), 88–96.
55. B. Hasselblatt and J. Schmeling, *Dimension product structure of hyperbolic sets*, in Modern Dynamical Systems and Applications, Cambridge University Press, Cambridge, 2004, pp. 331–345.
56. M. Hirayama, *An upper estimate of the Hausdorff dimension of stable sets*, Ergod. Theory Dyn. Syst. **24** (2004), 1109–1125.
57. V. Horita and M. Viana, *Hausdorff dimension of non-hyperbolic repellers. I. Maps with holes*, J. Stat. Phys. **105** (2001), 835–862.
58. V. Horita and M. Viana, *Hausdorff dimension for non-hyperbolic repellers. II. DA diffeomorphisms*, Discrete Contin. Dyn. Syst. **13** (2005), 1125–1152.
59. H. Hu, *Box dimensions and topological pressure for some expanding maps*, Commun. Math. Phys. **191** (1998), 397–407.
60. G. Iommi, *Multifractal analysis for countable Markov shifts*, Ergod. Theory Dyn. Syst. **25** (2005), 1881–1907.
61. T. Jordan and K. Simon, *Multifractal analysis of Birkhoff averages for some self-affine IFS*, Dyn. Syst. **22** (2007), 469–483.
62. A. Katok and B. Hasselblatt, *Introduction to the Modern Theory of Dynamical Systems*, Encyclopedia of Mathematics and Its Applications 54, Cambridge University Press, Cambridge, 1995.
63. G. Keller, *Equilibrium States in Ergodic Theory*, London Mathematical Society Student Texts 42, Cambridge University Press, Cambridge, 1998.
64. R. Kenyon and Y. Peres, *Measures of full dimension on affine-invariant sets*, Ergod. Theory Dyn. Syst. **16** (1996), 307–323.
65. M. Kesseböhmer and B. Stratmann, *A multifractal formalism for growth rates and applications to geometrically finite Kleinian groups*, Ergod. Theory Dyn. Syst. **24** (2004), 141–170.
66. M. Kesseböhmer and B. Stratmann, *A multifractal analysis for Stern-Brocot intervals, continued fractions and Diophantine growth rates*, J. Reine Angew. Math. **605** (2007), 133–163.
67. F. Ledrappier and L.-S. Young, *The metric entropy of diffeomorphisms I. Characterization of measures satisfying Pesin's entropy formula*, Ann. Math. (2) **122** (1985), 509–539.
68. F. Ledrappier and L.-S. Young, *The metric entropy of diffeomorphisms II. Relations between entropy, exponents and dimension*, Ann. Math. (2) **122** (1985), 540–574.
69. A. Lopes, *The dimension spectrum of the maximal measure*, SIAM J. Math. Anal. **20** (1989), 1243–1254.
70. N. Luzia, *A variational principle for the dimension for a class of non-conformal repellers*, Ergod. Theory Dyn. Syst. **26** (2006), 821–845.

71. N. Luzia, *Measure of full dimension for some nonconformal repellers*, Discrete Contin. Dyn. Syst. **26** (2010), 291–302.
72. R. Mañé, *The Hausdorff dimension of horseshoes of diffeomorphisms of surfaces*, Bol. Soc. Bras. Mat. (N.S.) **20** (1990), 1–24.
73. R. Mauldin and M. Urbański, *Dimensions and measures in infinite iterated function systems*, Proc. Lond. Math. Soc. (3) **73** (1996), 105–154.
74. R. Mauldin and M. Urbański, *Conformal iterated function systems with applications to the geometry of continued fractions*, Trans. Am. Math. Soc. **351** (1999), 4995–5025.
75. R. Mauldin and M. Urbański, *Parabolic iterated function systems*, Ergod. Theory Dyn. Syst. **20** (2000), 1423–1447.
76. H. McCluskey and A. Manning, *Hausdorff dimension for horseshoes*, Ergod. Theory Dyn. Syst. **3** (1983), 251–260.
77. C. McMullen, *The Hausdorff dimension of general Sierpiński carpets*, Nagoya Math. J. **96** (1984), 1–9.
78. K. Nakaishi, *Multifractal formalism for some parabolic maps*, Ergod. Theory Dyn. Syst. **20** (2000), 843–857.
79. J. Palis and M. Viana, *On the continuity of Hausdorff dimension and limit capacity for horseshoes*, in Dynamical Systems (Valparaiso, 1986), edited by R. Bamón, R. Labarca and J. Palis, Lecture Notes in Mathematics 1331, Springer, Berlin, 1988, pp. 150–160.
80. W. Parry and M. Pollicott, *Zeta Functions and the Periodic Orbit Structure of Hyperbolic Dynamics*, Astérisque 187–188, 1990.
81. Ya. Pesin, *Dimension Theory in Dynamical Systems: Contemporary Views and Applications*, Chicago Lectures in Mathematics, Chicago University Press, Chicago, 1997.
82. Ya. Pesin and V. Sadovskaya, *Multifractal analysis of conformal axiom A flows*, Commun. Math. Phys. **216** (2001), 277–312.
83. Ya. Pesin and H. Weiss, *A multifractal analysis of equilibrium measures for conformal expanding maps and Markov Moran geometric constructions*, J. Stat. Phys. **86** (1997), 233–275.
84. Ya. Pesin and H. Weiss, *The multifractal analysis of Gibbs measures: motivation, mathematical foundation, and examples*, Chaos **7** (1997), 89–106.
85. M. Piacquadio and M. Rosen, *Multifractal spectrum of an experimental (video feedback) Farey tree*, J. Stat. Phys. **127** (2007), 783–804.
86. M. Pollicott and H. Weiss, *The dimensions of some self-affine limit sets in the plane and hyperbolic sets*, J. Stat. Phys. **77** (1994), 841–866.
87. M. Pollicott and H. Weiss, *Multifractal analysis of Lyapunov exponent for continued fraction and Manneville-Pomeau transformations and applications to Diophantine approximation*, Commun. Math. Phys. **207** (1999), 145–171.
88. F. Przytycki and M. Urbański, *On the Hausdorff dimension of some fractal sets*, Stud. Math. **93** (1989), 155–186.
89. D. Rand, *The singularity spectrum $f(\alpha)$ for cookie-cutters*, Ergod. Theory Dyn. Syst. **9** (1989), 527–541.
90. M. Ratner, *Markov partitions for Anosov flows on n-dimensional manifolds*, Isr. J. Math. **15** (1973), 92–114.
91. D. Ruelle, *Statistical mechanics on a compact set with $\mathbb{Z}^\nu$ action satisfying expansiveness and specification*, Trans. Am. Math. Soc. **185** (1973), 237–251.
92. D. Ruelle, *Thermodynamic Formalism*, Encyclopedia of Mathematics and Its Applications 5, Addison-Wesley, Reading, 1978.
93. D. Ruelle, *Repellers for real analytic maps*, Ergod. Theory Dyn. Syst. **2** (1982), 99–107.
94. O. Sarig, *Thermodynamic formalism for countable Markov shifts*, Ergod. Theory Dyn. Syst. **19** (1999), 1565–1593.
95. S. Savchenko, *Special flows constructed from countable topological Markov chains*, Funct. Anal. Appl. **32** (1998), 32–41.
96. J. Schmeling, *Symbolic dynamics for β-shifts and self-normal numbers*, Ergod. Theory Dyn. Syst. **17** (1997), 675–694.

97. J. Schmeling, *On the completeness of multifractal spectra*, Ergod. Theory Dyn. Syst. **19** (1999), 1595–1616.
98. J. Schmeling, *Entropy preservation under Markov coding*, J. Stat. Phys. **104** (2001), 799–815.
99. K. Simon, *Hausdorff dimension for noninvertible maps*, Ergod. Theory Dyn. Syst. **13** (1993), 199–212.
100. K. Simon, *The Hausdorff dimension of the Smale–Williams solenoid with different contraction coefficients*, Proc. Am. Math. Soc. **125** (1997), 1221–1228.
101. K. Simon and B. Solomyak, *Hausdorff dimension for horseshoes in* $\mathbb{R}^3$, Ergod. Theory Dyn. Syst. **19** (1999), 1343–1363.
102. B. Solomyak, *Measure and dimension for some fractal families*, Math. Proc. Camb. Philos. Soc. **124** (1998), 531–546.
103. F. Takens, *Limit capacity and Hausdorff dimension of dynamically defined Cantor sets*, in Dynamical Systems (Valparaiso 1986), edited by R. Bamón, R. Labarca and J. Palis, Lecture Notes in Mathematics 1331, Springer, Berlin, 1988, 196–212.
104. M. Urbański and C. Wolf, *Ergodic theory of parabolic horseshoes*, Commun. Math. Phys. **281** (2008), 711–751.
105. P. Walters, *Equilibrium states for* β*-transformations and related transformations*, Math. Z. **159** (1978), 65–88.
106. P. Walters, *An Introduction to Ergodic Theory*, Graduate Texts in Mathematics 79, Springer, Berlin, 1982.
107. Y. Yayama, *Dimensions of compact invariant sets of some expanding maps*, Ergod. Theory Dyn. Syst. **29** (2009), 281–315.
108. L.-S. Young, *Dimension, entropy and Lyapunov exponents*, Ergod. Theory Dyn. Syst. **2** (1982), 109–124.
109. M. Yuri, *Multifractal analysis of weak Gibbs measures for intermittent systems*, Commun. Math. Phys. **230** (2002), 365–388.
110. Y. Zhang, *Dynamical upper bounds for Hausdorff dimension of invariant sets*, Ergod. Theory Dyn. Syst. **17** (1997), 739–756.

Index

B
base, 20
bounded variation, 30, 115
Bowen–Walters distance, 25
Bowen's equation, 2
box dimension, 45
    lower –, 46
    upper –, 46
BS-dimension, 43, 44
    spectrum, 127

C
coding map, 37
cohomologous functions, 20
cohomology, 19
    class, 20
conditional variational principle, 111, 115
conformal
    flow, 51
    map, 91

D
diameter, 45
dimension
    box –, 45
    Hausdorff –, 45, 46, 67
    lower box –, 45, 46
    lower pointwise –, 46, 82
    pointwise –, 61, 84
    spectrum, 127, 139, 141
        for the pointwise dimensions, 84, 92, 96
    upper box –, 45, 46
    upper pointwise –, 46, 82

E
entropy, 39
    local –, 62, 120
    lower local –, 120
    spectrum, 111, 124
        for the Birkhoff averages, 89, 105
        for the local entropies, 87, 120
        for the Lyapunov exponents, 121
    topological –, 40
    upper local –, 120
equilibrium measure, 41
ergodic
    decomposition, 47, 67
    measure, 40
expansive flow, 42

F
flow
    conformal –, 51
    expansive –, 42
    hyperbolic –, 33, 91
    suspension –, 19
    topologically mixing –, 42
    topologically transitive –, 107
function
    height –, 19, 25
    transfer –, 35

H
Hausdorff
    dimension, 45, 46, 67
    measure, 45
height function, 19, 25
horizontal segment, 25
hyperbolic
    flow, 33, 91
    set, 33

I
invariant measure, 40
irregular set, 87, 106

L. Barreira, *Dimension Theory of Hyperbolic Flows*,
Springer Monographs in Mathematics, DOI 10.1007/978-3-319-00548-5,

L
local entropy, 62, 120
locally maximal set, 34
lower
  box dimension, 45, 46
  local entropy, 120
  pointwise dimension, 46, 82

M
map
  coding –, 37
  conformal –, 91
  transfer –, 35
Markov
  chain, 36
  system, 35, 36
measure
  equilibrium –, 41
  ergodic –, 40
  Hausdorff –, 45
  invariant –, 40
  of maximal dimension, 70
multidimensional spectrum, 124, 127
multifractal analysis, 91, 127

P
pointwise dimension, 61, 84

R
rectangle, 35, 63
repeller, 91

S
segment
  horizontal –, 25
  vertical –, 26
semiflow, 25
  suspension –, 25
set
  hyperbolic –, 33
  irregular –, 87, 106
  locally maximal –, 34
spectrum
  BS-dimension –, 127
  dimension –, 84, 92, 96, 139, 141
  entropy –, 87, 89, 105, 111, 120, 121, 124
stable manifold, 34
suspension
  flow, 19
  over expanding map, 91
  semiflow, 25
symbolic dynamics, 36, 81

T
topological
  entropy, 40
  Markov chain, 36
  pressure, 39, 40
topologically
  mixing flow, 42
  transitive flow, 107
transfer
  function, 35
  map, 35
transition matrix, 36

U
u-dimension spectrum, 84
unstable manifold, 34
upper
  box dimension, 45, 46
  local entropy, 120
  pointwise dimension, 46, 82

V
variational principle, 111, 115
vertical segment, 26

Printed in the United States
By Bookmasters